I0473298

Design Thinking for Food and Beverage Innovation

Combining up-to-date theory and practice, this book provides a 360-degree overview of how to incorporate a holistic, pragmatic 'Design Thinking' inspired methodology into food and beverage product design. From the initial conceptual stages through to launch, it covers the key elements of both the design thinking methodology and product design, drawing on the author's personal experiences and diving deep into relevant industry case studies. It also offers focused insights into how empathy can build understanding of consumer motivations, and how to design product experiences that fit specific consumer desires and expectations. Ideal for industry professionals seeking to elevate their approach to product design, this book is a valuable resource for those passionate about food and beverages, eager to start their first or next innovation project.

Design Thinking for Food and Beverage Innovation

Alex Toft Nielsen

CRC Press
Taylor & Francis Group
Boca Raton London New York

CRC Press is an imprint of the
Taylor & Francis Group, an **informa** business

Designed cover image: Alex Toft Nielsen

First edition published 2026
by CRC Press
2385 NW Executive Center Drive, Suite 320, Boca Raton, FL 33431

and by CRC Press
4 Park Square, Milton Park, Abingdon, Oxon, OX14 4RN

CRC Press is an imprint of Taylor & Francis Group, LLC

© 2026 Taylor & Francis Group, LLC

ISBN: 9781041024644 (hbk)
ISBN: 9781041024620 (pbk)
ISBN: 9781003619352 (ebk)

DOI: 10.1201/9781003619352

Typeset in Times
by Newgen Publishing UK

Contents

SECTION II   *Understanding Design Thinking*

Preface

The idea to write this book was conceived while running in the summer of 2023. Running always seems to unleash creativity and spark new ideas for me. Sometimes it feels like each run is more of an ideation session than an exercise. For me, the perfect combination!

During the writing process, many runs helped shape the content, structure the narrative, and refine the innovation process you are about to read more about.

It has been my privilege to have worked with food and beverages in many ways since 1988, when I started as an apprentice chef. I was young, only 16 years old at that point, and it was a passion for the creativity of composing food that was my main motivation to start my education.

What I soon discovered was that the work as a chef is also hard work and it requires a wide range of skills to unleash the creativity.

Becoming a chef is like learning to play an instrument. You need to understand the basic theory and practice the hands-on techniques that bring the theory to life.

I also discovered the importance of the taste experience and how different people have different preferences.

I started to realize the importance of my audience. The guests in the restaurant.

Young chefs, and old ones too, can be somewhat like artists in their mindset! They compose the food to meet their own need for creativity and taste preferences.

The really successful ones hit the preferences of the local communities, and the ones that can attract an international audience have created something unique.

My path did not lead me to stay in the restaurant business. After a short time, I combined my curiosity about other cultures and desire to travel, and ended up cooking at the Danish Embassy in Kathmandu, Nepal. More precisely, teaching the Nepalese chef, Narayan, at the embassy how to cook French food.

It was a success in the sense that Narayan acquired some French cuisine skills, but I actually believe I learned far more from him than I was able to teach him. A completely new world opened up, and I was exposed to an assortment of new ingredients and cooking methods far removed from the French cuisine I was taught at the culinary school in Denmark.

But also the perspective on life that he gave me.

In Kathmandu, I had an important lesson in what empathy can provide.

I might have thought I was a hardworking chef in Denmark, but Narayan's long hours and sheer number of working days made me appreciate my own working conditions back home much more. By putting myself in his shoes, I gained a deeper understanding of his reality.

In my early twenties, I took a careerturn.

I went back to school to pursue a PBA in education, but also to start my first small company. It was essentially a private dining concept. I called it Art Culinaire – To emphasize that I was an artist in the culinary disciplines, I guess! It makes me laugh a bit to reflect upon the choice of name today.

What I did back in 1999 to achieve success wasn't offering a really interesting menu for customers to choose from, but rather tailoring a menu for their event, based on a personal interview with them.

I asked about their food preferences and what main ingredients they would like. I observed their choice of paintings on the walls and what kind of family they were. Did they like to travel, what their occupations were, how the kitchen looked, et cetera?

I think this was the first time I unconsciously applied design thinking in my work.

Today, my career has taken many twists and turns, but a common theme that has gained significance is the importance of understanding the audience you aim to reach.

This book is all about understanding.

It's about understanding the food and beverage industry and how to leverage innovation by applying a design thinking mindset.

It's been a privilege to to write it and deepen my own understanding in the process.

I hope you will enjoy reading it and find it useful.

My humble aim is that the content contributes to your understanding of the food and beverage industry and how design thinking can add value to the innovation and design of future products for consumers to enjoy.

Alex Toft Nielsen – *Aarhus, Denmark – Spring 2025*

Acknowledgements

OCEANS OF GRATITUDE AND LOVE TO MY WIFE AND FAMILY!

Writing this book has been great fun and a unique learning experience. Even though I am the sole author, I have been far from alone on the journey of getting the idea, scoping the content, formulating, and visualizing the book.

A lot of the knowledge I have obtained during my career and life has been a large contributor.

I would specifically like to thank:

All of the amazing colleagues I work with daily, and all my previous colleagues that I have worked with in the past.

My Indian spice connection, Vineesh Venugopal from Mevive International Trading Co, with whom I worked during the world's best curry powder project, and the global taste panel providing the product experience feedback.

Also, a big thanks to the founders and all the teachers behind the design thinking program "Food Architect" at IBC, and all the good people I have met during and after.

A heartfelt thanks to these inspiring individuals for their conversations, reflections, and feedback that helped to shape this book.

Leticia F. Garcia, Helle Dollerup, Julie Kjær-Madsen, Kirsten Møller Jensen, Phine Katrine Kjaer Wiborg, Emil Munck de Voss, Pernille Holm Vind, Eric Souza, Henrik Rewes, Jacob Klitgaard, Michael Brehm Midskov, Stephen M. Zollo, Fabian Apostoaie, Mads Thomsen, Idah, and the staff at Fadhila Cottages, Sulawesi.

And finally, to all of the authors of the literature I have read and everyone else whose input, in any form, contributed to this book.

Thank you!

Disclaimer

As I am not a native English speaker, ChatGPT has been a huge help to rephrase some of my words into more grammatically correct and fluent sentences. It has also been used to generate three images in Chapter 8 to bring the consumer personas to life, as one example of how AI can be used in the innovation process.

About the Author

Alex Toft Nielsen (Food Architect, PBA in Education, Certified Chef)
Alex's career journey is anything but conventional! Alex started his career with a passion for food and was a certified chef at the age of 20 already. With the mindset of an anthropologist, he has worked with teaching and learning about food in places like Kathmandu, Nepal; Kerala, India; and Sulawesi, Indonesia.

Since 2008, Alex has been deeply involved in innovation, working across both B2B and B2C in the food and beverage industry. His expertise spans over a decade of product development and food technology, including roles at major ingredient companies like **Döhler GmbH** in Germany. He also brings commercial experience as an Innovation Manager for Urtekram and Kung Markatta, two leading organic food brands in Denmark and Sweden, under the **Midsona Group**.

Throughout his career, Alex has contributed to numerous product launches across various markets and has navigated the challenges of unrealized projects, obtaining valuable hands-on lessons in the innovation process. In addition, he is the founder of the spice company **Aroma Spices** (founded in 2007 and sold in 2017), especially known for the World's best curry powder project and the award-winning product, CP44 Curry Powder.

Since completing the Food Architect program at IBC (International Business College, Denmark) in 2011, Alex has been a dedicated practitioner of design thinking methodologies, bringing these principles to life through his work in innovation. He has worked as a product designer, shaping and implementing the design thinking methodology at a global Food and Beverage FMCG company since 2020.

1 Introduction

WHY A BOOK ABOUT DESIGN THINKING FOR FOOD AND BEVERAGE INNOVATION?

Design thinking has become increasingly a more popular innovation tool across a range of industries. It has especially excelled in the IT industry.

But how to adapt it to the food and beverage (F&B) industry in a way that makes sense?

Back in 2011, I was first introduced to design thinking in a formal educational certificate program called "Food Architect,"[1] specially designed for the food and beverage industry. At that point in time, I worked as a product developer in a B2B company supplying liquid ingredient solutions, mainly targeting the dairy industry. On the side I was also the owner of a small spice company called Aroma Spices.

It was one of these famous turning points in a career where you have an AHA! moment.

This food architecture program had excellent instructors and an exciting trip to Istanbul, Turkey, designed to push us out of our comfort zones and engage in real ethnographic immersions. We were divided into small teams and had to decide on an area to explore and ideate solutions to fit. All targeted the food and beverage industry. The process culminated in a final presentation and an evaluation by the instructors.

The entire approach to innovation, by starting with an understanding of consumers and their lives, really resonated with me.

After the program, I was eager to learn more about the design thinking methodology but was surprised that I couldn't find any books on the topic within the food and beverage industry context.

I found a few F&B examples here and there, but mostly in the context of the user experience in a supermarket or online services helping consumers to live healthier. On the other hand, I found a lot of examples from the IT industry on how fast and easy prototyping and testing with consumers can be! The software industry is characterized by lightning-fast innovation and an overwhelming abundance of products. Software companies can iterate, test, and launch new offerings in weeks or months. The most obvious success stories displayed are those where the final product is an online solution, such as Airbnb or Uber.

DOI: 10.1201/9781003619352-1

1

This chapter was refined for grammar and fluency using ChatGPT-5.0.

At that point in time, I read what I could find on the topic and continued working hands-on with innovation and consumer centricity. However, a seed of an idea was planted!

The idea of writing a book about the design thinking methodology with relatable food and beverage examples to fit the reality of the industry.

If you are passionate about innovating new and exciting food and beverage products that consumers will love to buy, this book is for you.

You probably work in or with the food and beverage industry or aspire to do so.

It could be businesses like:

- Startup or small F&B companies with a limited range of products, fighting to achieve success.
- Medium-sized F&B companies with an established brand and category, aiming to expand their market share.
- Large players in the F&B industry with multiple product categories and brands, and a constant need to stay relevant and competitive.
- Business-to-business ingredient companies that support the industry with raw materials, and sometimes also insights.
- Retailers that launch products in their own brands – Aka, Private Label.
- Marketing agency business with the food and beverage industry as customers.
- Research companies, assisting the industry with insights.
- Other adjacent players in the F&B industry.
- Universities and other education systems that deliver F&B innovation as a part of the curriculum.

This book covers key elements of the design thinking methodology, and tools are drawn from various sources, combined with my personal experience of applying design thinking in the food and beverage industry.

Some of the commonly used terminology in design thinking may differ slightly, and the overall process and phases may go beyond what is typically defined as design thinking.

All of this is tailored to best suit food and beverage innovation and to provide a detailed journey through the multiple disciplines involved in F&B innovation.

A "Design Thinking" methodology adapted to fit Food and Beverage Innovation.

Design thinking is a methodology described in many ways in literature and is also constantly evolving.

Tim Brown, the author of "Change by Design" (2009), is the CEO of IDEO,[2] one of the pioneering design firms that popularized design thinking. He emphasizes human-centered design and innovation through empathy. His book[3] was revised and updated in 2019 and is considered an essential contribution to the design thinking methodology.

There are also many other valuable and insightful publications in the area. I recommend reading others as well, if you really want to deeply understand the mindset behind design thinking.

In design thinking, we always start with the problem, NEVER with the solution.

This mindset is one of the core pillars of traditional design thinking.

When you read the literature about the topic, it is mostly focused on radical innovation. Quotes from Steve Jobs and Albert Einstein indicate that innovation efforts should be placed in the task of identifying and understanding the problem, hence the product idea will come subsequently and be easy and fast to execute.

There are two types of innovation:[4]

1. Radical innovation is something new to the world and a very interesting area, but in food and beverage innovation, it is not so often the case. But it does happen! Often driven by R&D and new technological inventions.
2. Incremental innovation is built on existing product types, with minor or larger tweaks to differentiate what is already on the market.

Never start with the solutions …!

It is a great mindset to enable creativity with a starting point in a consumer need, but the risk is that it gets misinterpreted into a black box mindset by well-meaning facilitators.

Personally, I have experienced ideation workshops that turn into a competition of blue-sky ideas far from reality, resulting in no tangible outcome for the company. A waste of time and money.

Never start with the solution, seems the correct mindset, but it is a misconception to take it too literally.

Many of the existing food and beverage companies already have production lines and well-established product categories.

It makes no sense to imagine a far-from-reality setting where all types of products are in scope for an innovation project.

I believe that it is more than sensible to acknowledge that food and beverage innovation, is largely incremental within a set category. Some companies may operate across multiple categories, especially if they are not limited by their own production capabilities. It ultimately depends on the nature of the company.

Many good ideas already exist within a company and are constantly promoted by engaged employees, who are eager to have *given birth* to the next launch and success story.

It might be a great product idea that will solve real problems for grateful consumers, but then again, it might not.

My advice is to appreciate all ideas, wherever and whenever they pop up. That said, I am not recommending starting an innovation project based on a single product idea. That is rarely a good strategy but rather, collect the ideas and make a note of who the originators are.

Use them later in the process, where a proper foundation of consumer understanding is in place. This way, it is possible to validate whether the idea really fits a need in its original form, or if it should be adjusted, serve as inspiration for new ideas, or be discarded.

FOOD AND BEVERAGE EXAMPLES

Throughout this book, I will also use many concrete food and beverage examples to make design thinking more tangible for innovation in the F&B industry.

Some of them stem from real-life experience, while others were created specifically for this book.

The above-mentioned is some of the key elements I personally have been looking for when reading theory about design thinking, and all the adjacent frameworks and theories used in the industry.

A GUIDE BOOK

The aim is to provide the reader with detailed guidance on how to start an innovation project and show all the avenues of research and tools that may assist you in the journey.

A bit like a travel guide book: it will show good routes to travel and many options possible to pick, to enhance the quality of the trip.

When you travel, you won't be able to try all the recommended restaurants and hotels or go to the southern part of the country this time, but it will, nevertheless, be a great trip to remember.

This book will also present some company cultural travel advice as well: what to be mindful of and how to go about the curveballs that might come your way during the process.

I have never worked in an innovation project where hard choices of what to prioritize weren't the reality.

It is normal to feel that if you just had more time and resources to investigate more details about the "where and why" of a potential consumer demand, the ideas reflecting the opportunity would be possible to design even better.

The book is also a reminder to the CEOs of the F&B industry that successful product innovation is the foundation of a company's future, and embracing the mindset of design thinking is an investment with significant long-term returns. It is the way to secure product relevance and profitability in an ever-changing market.

The number of resources that can be allocated to implementing design thinking in innovation naturally depends on the type and size of the F&B company.

Startups and smaller companies will have to go with what is possible within their financial constraints. Larger corporations will mostly have larger budgets.

It is important to find the right balance and prioritize between innovating the right solutions for consumers or simply relying on the power of advertising.

Many companies, however, seem to mostly rely on the latter, looking at marketing budgets vs. innovation budgets.

My view is that it should go hand in hand.

It is of no use that you design the perfect product, and no one knows about it.

And what does it matter if you tell the world, in a remarkable way, that your company has a cool brand with great products, if no one likes the taste or notices any functional difference during and after consumption?

In fact, if done in collaboration, all the knowledge about consumer behavior, attitudes, needs, and motivation that can be harvested by using design thinking methodologies will inform how to design the product and how to let consumers know we are here, and we get you!

Danielle Sarver Coombs precisely frames the consumer perspective on brands and their products, in her brilliant consumer insight handbook:

The brands get you, so you get the brand.[5]

Throughout the course of this book, we will investigate many interesting food and beverage topics. It is not just an innovation process manual, but also a look into the "machine room" of the roles and expertise that sit across the different fields throughout the process.

This book will cover: the importance of creating delicious **tasting** products, how to anticipate consumer **future scenarios**, build a **strategy**, lead and **manage** a project, conduct and utilize **consumer** and business **research**, identify **demands**, translate them into **opportunities**, create **empathy** and enable **creativity**, unleash it in an **ideation** session, choose the **best ideas**, decide on the **fidelity** for a **prototype**, **test** the prototypes with **consumers**, go back and **iterate**, create **packaging design**, initiate product **development**, ensure **shelf life**, upscale, **validate**, get listings, build and launch **marketing** campaigns, **launch** the product, evaluate and **learn**.

As mentioned earlier, I hope that you will enjoy reading the book and will find it useful.

Chapters 2 to 5 set the stage by introducing some of the basics of the food and beverage industry.

Chapter 6 is an introduction to the design thinking mindset and methodology.

The following chapters, from 7 to 15, describe each phase of the design thinking innovation process, adapted to fit the F&B industry.

I recommend reading the entire book, as it is written as a connected story with many references to previous chapters and methodologies.

Read it at your convenience, and if you are working in the F&B industry and about to start an innovation project, use the book as a reference guide when you are planning the new project.

Whenever you encounter challenges or need clarification on specific aspects of the process, refer to relevant sections of the book for guidance.

Let's start the journey and begin with the essence of the industry – the food and beverages – we are passionate about innovating, producing, and delighting our consumers.

NOTES

1   https://kurser.ibc.dk/food-architect
2   **IDEO** is a global design and innovation firm that popularized **design thinking**, focusing on human-centered solutions. Known for designing the first Apple mouse, IDEO works across industries, emphasizing empathy, prototyping, and iterative problem-solving. www.ideou.com

3   *Change by Design, Revised and Updated: How Design Thinking Creates New Alternatives for Business and Society*, Tim Brown & Barry Katz, Harper Business, New York, 2019.

4   *Design-Driven Innovation: Changing the Rules of Competition by Radically Innovating What Things Mean*, Roberto Verganti, Harvard Business Press, Boston, MA, 2009.

5   *The Consumer Insights Handbook: Unlocking Audience Research Methods*, Danielle Sarver Coombs, Rowman & Littlefield Publishers, London, UK, 2021. Page 1.

Section I

Understanding the F&B Industry

2 Food and Beverages

Why do we live? A big question with many possible answers! Without getting too philosophical about it, we live because we are given life, and we are personally responsible for the value that we add to it. Why do we eat and drink? Another interesting question!

This topic can again have a multitude of answers, but in essence, we eat and drink, of course to stay alive. It is the fuel our body needs to function. If we don't eat, we will die. No way of avoiding it. It is mandatory.

So why not enjoy it?

I guess this is the question that previous generations, going back to an ancient point in time, have asked themselves. Luckily, our body is equipped with multi-sensory abilities. We have a very advanced mouth and nasal cavity, which allows us to perceive the flavors of the food we eat and the beverages we drink. It is a complex and integrated process that involves not only taste but also smell, texture, viscosity, temperature, visual cues, memories, and much more.

The brain combines all these sensory inputs to create our overall experience of flavor and enjoyment of the food. The enjoyment of food and beverages is the backbone in most cultures. Food and beverage products come in many shapes and forms. They are consumed on different occasions and with different motivations.

The focus of this book is on food and beverages that consumers buy in retail and use in various ways, that is as a convenient meal, as a snack, as refreshments, as an ingredient to prepare a dish, and so on. In retail, food and beverages can basically be divided by storage conditions.

There are three main storage categories:

1. Ambient (room temperature)
2. Chilled
3. Frozen

To give you an idea of the retail landscape, here are some examples from three different storage categories, including typical products and how they might be used.

DOI: 10.1201/9781003619352-3

This chapter was refined for grammar and fluency using ChatGPT-5.0.

AMBIENT

The ambient category includes some of the easiest products to handle, ship, and store. There is no need for refrigeration. These items offer convenience in both transportation and storage. The shelf life varies, depending on the product type, but it is generally long, making them ideal for both everyday consumption and for stocking the **pantry**.

Within this storage category, products can serve different functions. Some are raw ingredients for cooking, some are ready to consume, and others like ketchup, can be both an individual product and an ingredient in cooking. The category spans everything from pantry staples to indulgent treats, offering flexibility in how they are used, stored, and enjoyed. To get you into the feeling, we will take a short trip around the ambient aisles and reflect upon why we would buy them.

Beans and Pulses include dried beans, lentils, and peas, perfect for cooking from scratch using healthy, delicious plant-based protein for your evening dinner. **Bread** comes in a variety of types, from light to dark, full of seeds or plain, all of them catering to different occasions. Each country and even regions have its own traditional signature breads. It is the same for **Cakes and Cookies,** which offer precious moments of indulgence. Maybe you are expecting guests later that day, or always stock up as a sweet treat for the unexpected visit.

Condiments like ketchup, mustard, mayonnaise, and salad dressings provide that little extra touch to a meal or a salad. A meal can be cooked from scratch, but if you are in a hurry, **Instant Meals**, as instant noodles and meal kits offer quick preparation. It could also be that you are planning a hike and need food that is easy to carry.

Canned or jarred food can include all kinds of fish, meat, vegetables, and even ready meals. The taste experience of products like canned fish often surprises with its better-than-expected quality. Specifically, canned sardines have seen a growth in popularity. The trend is partly driven by social media and the influence of celebrity chefs, which have elevated the status of tinned fish from a humble pantry item to a gourmet delicacy. It is often served in wine bars.

Staples like **Oil or Vinegar**, range from premium olive, more budget-friendly sunflower oil to different brands of vinegar, which are all essential for cooking from scratch.

Similarly, **Pasta, Rice, and Grains** are available in many shapes and varieties. They are the foundation of countless meals around the globe.

Herbs and Spices, whether it is single spices like black pepper or blended seasonings, all add layers of flavor to dishes, desserts, cakes, and beverages.

The category also includes **Beverages,** such as soft drinks, juices, UHT milk, and chocolate milk. While these are stored ambient, they are often the best enjoyed when chilled. **Coffee and Tea** are available as beans, ground coffee, pods, or tea in different formats. In my house, it is an emergency if we run out of the coffee beans from the local roastery, or the unique product of Japanese Matcha Tea!

Lastly, on this fast and far from finished *tour de ambient*, the weekend is approaching, and good friends will be popping over, so consider whether a trip down the **Wine, Beer, and Alcohol** aisle is needed.

CHILLED

The chilled storage category is the most challenging to handle, ship, and store. Unlike ambient products, chilled items require controlled temperatures to maintain their freshness and safety. With a focus on freshness and convenience, the chilled category spans essential staples, meal solutions, and indulgent treats. All require careful handling to maintain their quality. Although the shelf life varies depending on the product, it is generally short.

Dairy products like milk, cream, yogurt, cheese, and butter, come in many forms and serve a wide range of purposes, from everyday staples to specialty ingredients.

The **fish and Seafood** category is present here with many retail products: raw, brined, pickled, or salted and smoked – what about a delicious smoked salmon for your salad, shrimps on toast, or the Danish classic: A traditional piece of S*mørrebrød*[1] with pickled herring for the lunch table?

Similarly, **vegetables and fruit** such as fresh cut fruit, salad mixes, and pre-cut vegetables provide convenience for snacking, salads, and meal preparation.

Meat products normally have a lot of space in the chilled section. These products range from fresh cuts of beef, chicken, pork, and lamb to marinated, ready-to-cook options, deli products, bacon, and sausages.

Ready-to-eat meals, including pre-packaged salads, sandwiches, pasta dishes, and sushi, all offer quick, convenient meal solutions for busy lifestyles.

Sauces and dips, such as fresh salsa, hummus, guacamole, and other freshly made condiments that add flavor and texture, whether used as dips, toppings, or cooking ingredients, are very convenient.

Juices and smoothies, from the classic orange juice to colorful smoothies, are essential. The refreshing properties and associated health benefits are embedded in consumers' minds. These products are, of course, also to be found in the ambient aisles, but if you are looking for the premium versions, the chilled aisle is the best place to look.

Finally, the **Desserts** in the chilled section include a variety of mouth-watering options such as puddings, mousses, cheesecakes, rice pudding, or crème brûlée. Yes – indulgence comes in many shapes and forms!

FROZEN

The frozen storage category strikes a balance between ease of handling, shipping, and storage. With versatility across meal occasions, frozen products combine convenience with quality, making them an essential part of modern food storage and preparation.

With a long shelf life and minimal waste, frozen products offer convenience without compromising too much on quality. Some items serve as ingredients for home cooking, while others are ready to eat or require minimal preparation.

Ice cream is one of those products you will find in the frozen section. This is definitely a must stop, right? Alongside ice cream, you will also find a large range of other products, also to be found in the chilled aisle, now in a frozen version.

Frozen vegetables and fruit, such as peas, berries, and spinach, are perfect for smoothies, side dishes, and baking.

Meat products, including frozen chicken breasts and beef patties, provide meal preparation flexibility and convenience for cooking.

Similarly, products such as frozen salmon, cod fillets, or tiger prawns offer healthy **seafood,** normally at a cheaper price compared to the chilled version.

Bread and cakes of all kinds are, of course, also available frozen, but some types of dough, such as croissants or puff pastry, are only available in frozen form.

Ready meals like frozen lasagna, pizza, shepherd's pie, butter chicken, or Thai green curry are great choices to stock up on in the freezer. And despite the usual assumptions, frozen ready meals often deliver a better product experience than their chilled counterparts.

You might wonder why that is. Frozen meals are often flash frozen shortly after cooking or assembling, locking in freshness, flavor, and nutrients. Since freezing naturally extends shelf life, frozen meals don't need additives or preservatives as used in chilled products.

THE ASSORTMENT

All products have a unique category and unit codes. Each product belongs to a specific product category, which groups products based on shared characteristics such as product type, usage occasion, or target consumer. Retailers and producers do not necessarily use categories the same way.

What remains consistent across the industry is the following: *stock keeping unit codes* are always referred to as **SKUs**. They reflect the total portfolio of what is registered as products that are going to be produced, are produced, or have been produced. The more SKUs you must handle in your assortment as an F&B company, the more complicated and time-consuming it gets. Some of the SKUs might not be very successful in terms of the profits they provide. In most companies, it is actually a few percent of the SKUs that bring a significant part of the earnings.

An exercise that is often used to look at this issue is the SKU Tail exercise. It is, however, complicated, as there are many things to consider.

Are some of the low-performing products important for your company in other ways?

A product could be very important for one of your customers, and to keep the good relationship, it will be kept within the assortment, even if it is not making a profit. But the total commitment with the company includes some of the real profitable products as well. It is a dilemma that cannot be solved with a formula.

F&B INNOVATIONS: PAST AND FUTURE

Impactful Innovations from the Past

When you look back in time at the development of the food and beverage area, many innovations have reached the global market. Some of them are really changing the way we, as a species, consume our nutrition. We are the largest predator on the planet, and on average, we will eat mostly anything we find delicious, nutritious, and safe to consume. Technology has delivered numerous groundbreaking innovations over the past two centuries.

The three overarching themes are food safety, shelf life, and convenience. But also nutrition, health, and functionality have been major drivers for innovation. The design thinking methodology, as a structured methodology, is from the late 20th century and hence not used in many of these innovations, but the mindset of solving problems for the consumers certainly was, along with the innovator's aspirations in the companies, pursuing growth and success.

Some of the impactful and interesting innovations are the following:

Canned Food

Developed in the early 19th century, canned food revolutionized shelf life and food storage, making it possible to preserve meals for long periods without refrigeration. An innovative breakthrough in both food safety and convenience!

Pasteurization

Pasteurization[2] is a heat treatment process developed by the French chemist, Louis Pasteur, in the 1860s. Originally invented to prevent spoilage in wine, it is now widely used in the dairy industry and across the broader F&B industry to enhance safety and extend the shelf life.

Carbonated Soft Drinks

Originally developed for medicinal purposes, Coca-Cola[3] began in the late 1800s as a medicinal tonic, originally sold in pharmacies. However, carbonated soft drinks grew into a global category. The category's success reflects not just taste appeal, but also advancements in the processing, bottling, and branding.

Powdered Baby Formula

A critical innovation for infant nutrition was the powdered baby formula. It provided an accessible alternative to breastfeeding, especially important for working parents or those without access to maternal milk.

Instant Coffee

By turning brewed coffee into a powder, instant coffee offered unprecedented convenience and became a staple in homes globally. The taste experience has improved over the years, but instant coffee is clearly one of the examples where convenience beats taste.

Frozen Foods

The freezing technology made it possible to preserve meals while maintaining flavor and nutrition. This innovation redefined home cooking and made seasonal foods available all year round.

Processed Cheese

In 1916, James L. Kraft patented[4] a method for pasteurizing cheese, enabling the production of shelf-stable cheese that did not require refrigeration. This innovation led to the creation of Kraft's processed cheese, which gained widespread popularity.

Quorn – The Mycoprotein Pioneer

Quorn's introduction of mycoprotein[5] in the 1980s was a groundbreaking innovation. Mycoprotein is a food ingredient high in protein and fiber made by fermenting the fungus, *Fusarium venenatum*.

Quorn's mycoprotein initially stayed under the radar, most likely due to limited market readiness at the time. Ironically, Quorn's foundational patents, filed in 1985, expired in 2010 – just as global interest in sustainable and alternative proteins began to take off.

The timing meant that other companies could legally begin to produce mycoprotein, using the previously patented processes, as long as they used different brand names.

Instant Noodles

Affordable, fast, and satisfying — instant noodles transformed the global eating habits, especially in urban and student settings. Instant noodles were invented by Momofuku Ando[6] from Japan.

Tetra Pak Packaging

This breakthrough in aseptic packaging[7] extended the shelf life of easily spoiled liquids like milk and juice without refrigeration, while also offering lightweight, easy-to-store packaging.

Microwaveable Meals

Paired with the rise of the microwave oven, these meals brought unmatched speed to the table, reflecting the shifting lifestyles and the need for microwaveable meal solutions.

Artificial Sweeteners

Designed for those seeking low- or no-calorie alternatives to sugar, artificial sweeteners addressed both health concerns and dietary needs. Simultaneously, they have also sparked a lot of debate about natural vs. artificial products.

Energy Drinks

This is a newer category, driven by functionality. Energy drinks often offer a quick caffeine boost and other active ingredients. Their popularity highlights changing consumption patterns, especially among younger demographics.

Surely, this list could go on, as many impactful F&B innovations have already seen the light of day, but let's move from the past and turn to what the future might bring.

Radical Innovation

Radical innovation refers to something entirely new to the world. It may involve the combining of existing raw materials in a new way, for example, in creating a new functionality or benefit. Sometimes it solves a tension that the existing products on the market have not yet been able to address.

In many cases, radical ideas are not feasible within the current production capabilities, and they will require additional research and process design. They will also likely demand investments in new production equipment. In some cases, even new legislation might be required to allow for human consumption.

Securing Intellectual Property

Securing the rights to intellectual property (IP) can be key to the success of F&B companies, therefore, applying for patents should be prioritized. It is also worth monitoring new patent applications, both to ensure that you are not at risk of violating a pending patent, but also because they can serve as a valuable inspiration for new projects.

"Inside-out" innovation is often used to describe the opposite of consumer-centered innovation. These projects focus on new processes, novel technologies, and exciting ways to transform known ingredients into new tastes, textures, or even features beyond imagination. The risk is that, while the idea might be very interesting from a science and technology perspective, it may not be interesting or useful to any consumers on the market. To reduce that risk, it is important to apply design thinking, even in innovation projects that do not start by identifying a consumer problem.

Rethinking Protein for a Sustainable Future

With some of the radical innovations of the past in mind, what might the future look like? In recent years, the accelerating sustainability crisis has motivated innovation efforts aimed at creating products with a smaller climate footprint. Historically, planet and human-friendly food and beverages were developed and marketed as less polluting and healthier to consume, as their production avoided the use of chemicals to enhance yield.

The organic food movement, which emerged in the early to mid-20th century as a response to industrialized agriculture, emphasized natural farming methods without using synthetic pesticides, fertilizers, or genetically modified organisms (GMOs). These efforts were often framed around health, ethical consumption, and environmental awareness. Over the past two decades, particularly since 2010, sustainability and the climate impact of CO_2 emissions have become central drivers of innovation. Especially the high CO_2 footprint of meat production, particularly that of cows, has nudged established companies and startups to initiate innovation projects around alternative proteins. Many alternative proteins to meat have been explored, developed, and used for centuries. Soybeans are one of the oldest and most efficient alternatives. Tempeh, a fermented soybean product, has been a staple in Indonesian cuisine for several centuries, offering a protein-rich, plant-based meat alternative long before the current sustainability discourse began.

In recent decades, the interest in alternative proteins has grown. This was driven by climate change and the global need to feed a growing population. In the book *How to Avoid a Climate Disaster*,[8] Bill Gates emphasizes the urgency of reducing emissions from food production, which accounts for about a quarter of the global greenhouse gases. This means rethinking not only what we eat, but also how we grow and produce it.

Many companies responded to the trend and began innovating. Plant-based products like those from *Beyond Meat*[9] and *Impossible Foods*,[10] using peas, soy, and chickpeas to mimic meat, started to emerge in the early 2010s. Despite all the innovation, purpose, and hype, companies like Beyond Meat have struggled in recent years.[11] After a strong start, the company has seen a decline in sales and interest, especially in its main market.

Fewer people are buying plant-based meat than expected, partly because of high prices, strong competition, and concerns about the taste and health. This shows that making sustainable food is not just about new technology alone. It also needs to meet everyday needs like affordability and a great product experience. As the plant-based category matures, a new wave of products is emerging. Products that don't try to mimic meat, but instead embrace simpler ingredients, minimal processing, and a more natural approach.

On the flip side, there are also a lot of initiatives in a more artificial direction. Especially two technologies are being explored and developed:

1. Cultivated meat, grown from animal cells by companies like *Upside Foods*[12] and *GOOD Meat*.[13]
2. Precision fermentation, used by *Perfect Day*[14] to create dairy proteins without animals.

Although these technologies promise environmental benefits, they, however, also raise questions around consumer acceptance, especially when they involve genetic modification (GMO). While GMOs have long been viewed with caution, attitudes may shift if such methods offer clearly lower climate footprints. This is a shift that will need to originate from both consumers accepting the new technology and authorities providing room in legislation to allow it. Dilemmas with new technology and what is considered harmful to consume, are old and very familiar situations in societies.

Growing up, I certainly remember people worried about microwave technology and how it could potentially be harmful. I guess it is a normal human reaction to be afraid of new inventions, if they are hard to understand, or seem too good to be true. The microwave prevailed, however, and is standard kitchen equipment in many homes around the globe. The F&B industry has supported development by introducing countless new microwaveable products and continues to do so.

The Process Puzzle

Simultaneously, there is also a counter movement against what is considered processed food, and why particularly to avoid it. Ultra-processed food should be avoided.

Tim Spector,[15] a prominent expert on nutrition and gut health, has highlighted concerns about ultra-processed foods (UPFs). Many innovations from the past have certainly introduced more processing into the food and beverage products available to consumers, compared to how they were produced before the industrial revolution. What is viewed as natural, and to what level a process can be classified, is a complex area, therefore I will refer to the NOVA[16] definition for further information:

The NOVA classification categorizes foods based on their level of processing into four groups:

1. **Unprocessed or Minimally Processed Foods**
2. **Processed Culinary Ingredients**
3. **Processed Foods**
4. **Ultra-Processed Foods**

Nothing is a Constant!

When I started to write this book, the main topic discussed in the F&B industry was sustainability, a very important topic that influenced company strategies and the scope of innovation projects. That said, the state of the world is constantly evolving, and what seems super important today might lose importance or focus tomorrow.

Currently, the topic of war seems to drown out the sustainability agenda, which is, of course, very sad. War over land and trade will, in the same way, also be reflected in the new strategies and innovation projects of F&B companies. Hopefully, a trend that will pass quickly …

NOTES

1 *Smørrebrød* is a traditional Danish open-faced sandwich. The word literally means "buttered bread," but it's much more than that. It is a beautifully crafted slice of rye bread topped with a variety of ingredients. The main toppings can range from cold cuts, pickled herring, liver pâté, smoked salmon, eggs, shrimp, or potatoes. Garnishes can be onions, capers, fresh herbs, remoulade, mayonnaise, et cetera. It is often eaten with a knife and fork, especially at lunch or during special gatherings.
2 www.britannica.com/biography/Louis-Pasteur
3 www.coca-colacompany.com/about-us/history
4 www.britannica.com/money/Kraft-Heinz
5 www.quorn.co.uk/about-quorn
6 www.cupnoodles-museum.jp/en/osaka_ikeda/about/#about_momofuku_heading
7 www.tetrapak.com/en-us/about-tetra-pak/who-we-are/heritage?
8 *How to Avoid a Climate Disaster: The Solutions We Have and the Breakthroughs We Need*, Bill Gates, New York, Knopf, 2021.
9 www.beyondmeat.com/
10 https://impossiblefoods.com/
11 www.wired.com/story/beyond-plant-based-meat-sales-trends-us-europe/
12 https://upsidefoods.com/

13 www.goodmeat.co/
14 https://perfectday.com/
15 https://zoe.com/learn/podcast-ultra-processed-food
16 www.researchgate.net/publication/331086204_Ultra-processed_foods_what_they_are_
     and_how_to_identify_them

3 Nutrition Science

Nutrition Science is the study of how food and its nutrients affect the body. It explores how the body digests food and beverages, absorbs nutrients, uses them for energy, growth, and essential bodily functions such as maintaining tissues and supporting the immune system.

In the food and beverage industry, a solid understanding of nutrition is essential and plays a critical role in product design. Considerations of nutritional composition influence everything from ingredient selection and the portion sizes, to claims on packaging and consumer messaging. A common way to understand and communicate nutrition to consumers is by grouping nutrients into two main categories namely, macronutrients and micronutrients.

MACRONUTRIENTS

Macronutrients are nutrients the body needs in large amounts to provide energy and to support essential functions like growth, repair, and movement. These macronutrients include carbohydrates, proteins and fats. These should be specified in the nutrition declaration on the packaging. The energy they supply is typically highlighted on the food and beverage packaging, shown as kilojoules (kJ) and kilocalories (kcal), helping consumers to quickly see how much "fuel" the product provides.

CARBOHYDRATES

Carbohydrates are compounds found in foods that consist of single or multiple sugar molecules in various forms. They serve as the body's main source of readily available energy, and they also include dietary fiber, which supports digestive health. Common sources include fruits, sugar, starchy vegetables, and whole-grain products.

PROTEINS

Proteins are compounds in foods made up of chains of amino acids, which the body uses to build and maintain tissues such as muscles, as well as vital components like

DOI: 10.1201/9781003619352-4

19

enzymes and red blood cells. Protein can also serve as a source of energy. Common sources include animal products, beans, pulses, nuts, and seeds.

FATS

Fats and oils are concentrated sources of energy and play several important roles in the body. They provide energy, support cell function, and aid in the absorption of vitamins. Common sources include a variety of oils (such as olive, rapeseed, and sunflower), butter, fatty fish, nuts, seeds, avocados, and full-fat dairy products such as milk (around 3.5% fat), cheese, and yogurt.

MICRONUTRIENTS

These nutrients are required in smaller amounts, but they are essential for overall health and development. Many important vitamins and minerals contribute to the vital body functions, although they are not always listed in the nutrition declaration on the food packaging.

VITAMINS

Four of the most well-known vitamins and their key roles are the following:

- **Vitamin C** – Supports immunity and skin health and is naturally found in citrus fruits.
- **Vitamin D** – Helps to maintain strong bones. It is naturally found in fatty fish and is produced by the body through exposure to sunlight.
- **Vitamin A** – Important for vision, immune function, and skin health. It is naturally found in carrots, sweet potatoes, spinach, and dairy products.
- **Vitamin B12** – Supports nerve function and red blood cell production. It is naturally found in meat, fish, and dairy products.

MINERALS

Five of the most well-known minerals and their key roles are the following:

- **Calcium** – It is essential for strong bones and teeth. It is naturally found in dairy products and leafy greens.
- **Iron** – It supports the production of red blood cells and helps to carry oxygen throughout the body. It is naturally found in red meat, beans, lentils, and spinach.
- **Potassium** – It helps to regulate the fluid balance, muscle contractions, and nerve signals. Potassium is naturally found in bananas, potatoes, and legumes.
- **Magnesium** – It supports numerous biochemical reactions in the body, contributing to muscle and nerve function. It is found naturally in nuts, seeds, whole grains, and leafy greens.

- **Sodium (salt)** is also an essential mineral, but it is a special case. While the body needs only a small amount to help regulate the fluid balance, nerve signals, and muscle function, many people consume too much. This can increase the risk of high blood pressure.

NUTRITION DECLARATIONS

Let's look at what the consumers are presented with on pack.

As an example, a butter chicken ready meal from a UK supermarket.

TABLE 3.1
Butter Chicken With Rice

Ingredients: Cooked Rice [Water, Basmati Rice, Rapeseed Oil, Salt, Concentrated Lemon Juice, Cumin Seeds, Cardamom Pods, Cardamom Powder, Bay Leaf Powder, Colour (Curcumin)], Cooked Marinated Chicken Breast (21%) [Chicken Breast, Tomato Purée, Ginger Purée, Garlic Purée, Cornflour, Salt, Soya Oil, Yogurt Powder (Milk), Green Chilli Purée, Water, Palm Oil, Chilli Powder, Yogurt (Milk), Skimmed Milk, Coriander Powder, Cumin Powder, Colour (Paprika Extract), Ginger Powder, Cinnamon, Black Pepper, Mace, Fenugreek, Star Anise, Turmeric, Basil], Single Cream (Milk), Tomato Purée, Tomato, Onion, Ginger Purée, Garlic Purée, Butter (Milk), Cornflour, Tomato Juice, Honey, Coriander Powder, Paprika, Cumin Powder, Salt, Green Chilli Purée, Fenugreek, Turmeric, Cinnamon, Colour (Paprika Extract), Cardamom Powder, Clove, Fennel, Dill.

Values	Per 100g
Energy	613kJ / 146kcal
Fat	6.1g
of which saturates	2.9g
Carbohydrate	14.0g
of which sugars	1.8g
Fibre	1.6g
Protein	8.0g
Salt	0.43g

The Butter Chicken ready meal provides detailed information on the total amount of calories and the macronutrients on its packaging. However, specific details about vitamins and minerals (micronutrients) are often not included. This omission is common in food labeling, as regulations typically only require macronutrient information, unless a product is fortified or makes specific health claims related to certain micronutrients.

Meal replacement products like a *Meal in a bottle* are an example of products where the full declaration is needed. That will be presented like the following:

TABLE 3.2
Oat & Chocolate Meal Shake

Ingredients: Oat Milk (Water, Gluten Free Oats), Coconut Cream (6%), Banana (6%), Soya Protein, Dates (1.7%), Pure Hazelnut Butter (1.3%), Cocoa Powder (1.2%), Vitamin & Mineral Blend, Cornstarch, Concentrated Monk Fruit Infusion, Citrus Fibre, Natural Flavorings, Vitamin & Mineral Blend: Minerals (Chloride, Potassium, Phosphorus, Magnesium, Calcium, Iron, Zinc, Manganese, Copper, Molybdenum, Chromium, Lodine, Selenium), Vitamins (a, K, Pantothenic Acid, E, C, Niacin, D, B6, Riboflavin, Thiamin, Biotin, Folate, B12), Choline

Values	Per 100g	Values	Per 100g
Energy	**266kJ / 64kcal**		
Fat (g)	2.3	**Folic Acid (μg)**	5.00
of which saturates (g)	1.1	**Vitamin B12 (μg)**	0.19
Carbohydrate (g)	4.7	**Biotin (μg)**	5.4
of which sugars (g)	2.3	**Pantothenate (mg)**	0.58
Fiber (g)	1.0	**Chloride (mg)**	1180
Protein (g)	5.1	**Calcium (mg)**	33.5
Salt (g)	0.30	**Potassium (mg)**	307
Vitamins & Minerals		**Phosphorus (mg)**	62
Vitamin A (μg)	60.6	**Magnesium (mg)**	45.5
Vitamin D (μg)	0.45	**Iron (mg)**	1.2
Vitamin E (mg)	1.6	**Zinc (mg)**	0.69
Vitamin K (μg)	4.7	**Copper (mg)**	0.28
Vitamin C (mg)	11	**Manganese (mg)**	0.55
Thiamin (mg)	0.10	**Selenium (μg)**	4.4
Riboflavin (mg)	0.25	**Chromium (μg)**	16
Niacin (mg)	1.3	**Molybdenum (μg)**	20
Vitamin B6 (mg)	0.35	**Iodine (μg)**	9.60

FUNCTIONAL FOOD AND BEVERAGES

Functional foods and beverages are products designed to provide health benefits beyond basic nutrition, blurring the lines between the traditional F&B and the supplement industry. Functional foods and beverages can be added with ingredients like vitamins, minerals, probiotics, fiber, healthy fats, or botanical compounds. F&B products that support areas like gut health, immunity, or mental performance are well-known, and much innovation is focused on supporting healthy lifestyles. The innovation of these types of products reflects a broader consumer interest in using everyday food and drinks, not just for nourishment or pleasure, but to support the overall health and performance.

It's fair to say that innovating functional foods and beverages, and communicating their benefits, is a complex business. Health claims on F&B products are often heavily regulated to protect consumers from unproven or misleading promises, which is, of course, a good thing.

What makes it even more challenging, however, is that legislation varies across the globe. As an example, what is allowed in the US might not be permitted in Denmark. While this category has seen rapid growth in recent decades, the idea of food as a vehicle for wellness has deep historical roots. A pioneering example is Yakult[1], a fermented dairy drink developed in 1935 by the Japanese microbiologist, Dr. Minoru Shirota. The product was created to improve the consumer's gut health through a specific strain of probiotics. It is still on the global market to this day, which is very impressive!

One thing is how healthy and nutritious a product might be, but what is even more important to consumers is how it tastes, looks, and feels.

Let's investigate the extremely interesting field of how the product experience can be described and tested.

The field of sensory and consumer science.

NOTE

1   www.yakult.co.jp/english/company/history.html

4 Sensory and Consumer Science

THE CONSUMPTION EXPERIENCE

There is a strong link between sensory science and consumer science, often referred to collectively as sensory and consumer science (SCS). This tasty science is basically divided into 3 categories:[1]

1. *Discrimination tests that aim at detecting subtle differences between products.*
2. *Descriptive analysis (DA), also referred to as "sensory profiling," aims at providing both qualitative and quantitative information about a product's sensory properties.*
3. *Affective tests. This category includes hedonic tests that aim at measuring consumers' liking for the tested products or their preferences among a product set.*

The complexity of the consumption experience is one of the reasons why the general design thinking methodology comes a bit short in food and beverage innovation. The role of prototyping and feedback is more complex than in many other industries, especially in the IT industry.

The sensory properties play a crucial role in evaluating and describing the overall sensory experience of food and beverages.

ORGANOLEPTIC EVALUATION

Organoleptic properties[2] refer to the sensory characteristics or qualities of a substance that can be perceived by the senses. Organoleptic evaluation is the measurement of the sensory qualities of a product. It can be described as the sensory foundation within SCS, where the data about product properties meets the consumer attitudes, behavior, and preference insights.

The key components of the five organoleptic properties are the following:

1. Taste
2. Smell
3. Sight

DOI: 10.1201/9781003619352-5
This chapter was refined for grammar and fluency using ChatGPT-5.0.

4. Touch
5. Hear

Taste

The five scientifically recognized basic tastes are the following:

1. Sweet
2. Salty
3. Sour
4. Bitter
5. Umami

In many textbooks, you will find a *tongue map* showing specific areas on the tongue where the basic tastes are perceived. However, as with many other aspects of life, science has later concluded that the idea of a tongue map is a misleading oversimplification of a more complex system of taste perception.[3]

Sweet

We learn to appreciate sweetness from the day we are born, as mother's milk is sweet, and growing up in most cultures often involves sweet treats as rewards and sources of comfort. Sweetness provides both indulgence and energy. Just think of biting into a ripe peach after a workout and the sensation it brings! Natural sweetness comes from sugars, providing a taste that feels universally satisfying.

There are also other ingredients, like artificial sweeteners, that can provide sweetness without delivering energy. Sugar is also widely used to preserve fruit and vegetables, e.g., jam, fruit syrup, and pickled vegetables. Even some meat and fish products are preserved with sugar.

Salty

Salt is essential to life and deeply ingrained in the human preference, much like sweetness. While sugar provides energy and fuels growth, salt serves as a cornerstone for survival, maintaining critical functions in the body. Salt is also enhancing the taste of the ingredients it is added to, and is a cornerstone in cooking food. From the simply satisfying taste of salt on a boiled egg to more complex dishes or sauces.

Salt is also widely used to preserve food, e.g. fish, meat, and vegetables. Salt is a naturally occurring mineral (NaCl), composed primarily of sodium chloride.

Sour

To some people, the sour taste may be more an acquired than naturally preferred taste, but it is nevertheless, equally important. In nature, the sour taste often indicates the presence of healthy acids and vitamins. Sour foods like citrus fruits (lemons, oranges, Yuzu, etc.) have a high content of vitamin C, which can boost the immune system. In fermented foods like kimchi or yogurt, the sour taste is also a trademark.

From a culinary point of view, the sour taste is a very important tool to balance a dish. Who doesn't recognize the effect of squeezing a lemon over seafood or sipping a glass of fresh lemonade on a hot day, offering a nice refreshing sensation?

Acidic products like vinegar are also a very important ingredient in preserving food, e.g. in pickled vegetables and fish.

Food Preservation

As alluded to, sweet, salty, and sour components have been used for generations to preserve food. This long-standing relationship with food preservation has made these taste sensations deeply fundamental to us, as they signal nourishment and safety, but also a connection to our culinary and cultural heritage.

Bitter

Unlike sweet, salty, and sour, bitterness is not associated with food preservation. Instead, its evolutionary role is to act as a natural warning system, helping humans to detect potentially toxic or spoiled substances. Bitter is an even more acquired taste, often appreciated more with age. Some of the classic steps toward appreciating bitter tastes begin, for example, when you start drinking coffee. You might not like the bitter taste, but you find it cozy and sociable to drink coffee. Often, people add sugar and milk to the coffee to balance the taste experience.

From a culinary point of view, bitterness is a taste to be handled with care. It can add complexity and depth to food and beverages but it is also a polarizing taste. Common ingredients that provide bitterness include coffee, tea, cocoa, hops, citrus peels, and some leafy greens.

In beer making, the bitterness is provided by the hops, and in some beer types, it is a trademark – especially in IPAs (India Pale Ales), which are known for their strong bitterness.

Umami

Umami,[4] often described as *savory*, is generally tied to foods rich in amino acid L-glutamate, as well as nucleotides. Without deep-diving into the chemical topic, as it is complex and not the aim of this book, it is particularly the amino acid L-glutamate that is important.

Monosodium glutamate (MSG) was invented in 1908 by Kikunae Ikeda, a Japanese scientist. While studying the savory flavor in kombu seaweed broth, Ikeda isolated glutamic acid as the source of the distinctive taste he later named *umami*. To make this flavor more accessible, he developed MSG as a stable and easy-to-use seasoning.[5] The umami taste can be found in natural ingredients such as tomatoes, shiitake mushrooms, and seaweed. However, it is even more prominent when it develops during chemical processes, like cooking and fermentation. Think of the 'meaty' flavor in a chicken soup, or the savory depth of soy sauce, beyond its salty taste. Regarding the preservation properties of sugar, salt, and acids, umami is, in many cases, the actual result of preservation. Especially in the case of fermentation. Familiar examples include soy sauce, miso, fish sauce, and Parmesan cheese.

When the world-famous Copenhagen-based restaurant, Noma, started their journey to create delicious tastes out of ingredients that are locally sourced in the Scandinavian terroir, a key enabler was fermentation. The Japanese tradition of using mold spores like Aspergillus Oryzae to ferment classics like Sake, Soy Sauce, and Miso is world-famous. The same fermentation principles utilized on Nordic ingredients resulted in something new, very umami, and very tasty!

SMELL

In organoleptic evaluation, smell plays a crucial role in flavor perception, particularly through retronasal olfaction, which occurs when volatile compounds from food or beverages travel from the mouth to the olfactory receptors in the nasal cavity. Before consumption, the initial perception of smell happens when we sniff or inhale the aroma of food. This prepares the brain for what we are about to taste.

As food is chewed or beverages are sipped, volatile compounds are released and travel up the throat into the nasal cavity. The brain integrates these retronasal odor signals with taste signals from the tongue, creating the full experience of flavor. This is why strawberries taste like strawberries, even though the tongue only detects sweetness and acidity.

This interaction explains why holding your nose while eating reduces flavor perception. It blocks retronasal olfaction. The aroma of a product contributes significantly to the overall flavor perception. Smell and taste together create what is commonly referred to as flavor. Compared to the basic tastes, the area of food aromas is very complex.

Aroma vs. Flavor

These two terms are important to understand, because they are often mixed up, and it is very confusing!

Aroma: It is primarily related to the sense of smell, detected by the nose, and involves volatile compounds, e.g. the aroma of lemon, vanilla, butter, biscuits, and so on.

Flavor: Flavor is extensively used in the same way as aroma, in the literature as well as in the F&B industry.

Flavor can also mean the multisensory experience that combines the basic tastes, aroma, mouthfeel, and other sensory inputs that create a complete perception of a food or beverage. The total flavor experience of a lemon cheesecake, e.g., is a harmonious blend of sweetness, tanginess, and richness. The creamy and smooth cheesecake filling melts in your mouth, balanced by the bright citrusy notes of fresh lemon juice and zest. The buttery, slightly crumbly biscuit base provides a satisfying contrast.

In essence, while aroma contributes significantly to flavor, flavor encompasses a broader range of sensory experiences, making it a more complex and comprehensive perception.

Flavors as an F&B Industry Ingredient

In the F&B industry, aromas are normally called *flavors*. While aroma remains a key contributor to flavor perception, this book will primarily use flavor to reflect the common usage in the food and beverage industry.

A Flavorist

There is a professional line of work in the flavor industry, called a flavorist. An F&B flavorist is a specialized professional who creates and develops flavors for food and beverages. Flavorists are typically employed by companies, usually referred to as flavor houses, to develop and produce flavors.

By using a mix of chemistry, sensory science, and creativity, they design natural and artificial flavors across all categories of ingredients and even dishes. Their expertise varies from fruit and berries to spices, chicken stock, and cheesecake. In essence, if you can imagine it, they can design it!

Categories of Flavors

On a food label in the European Union, flavorings must be declared according to the specific definitions and regulations set out by EU law.[6] An example here is the current options for declaring flavorings in the EU on a food label. In other parts of the world, it will be different.

N.B. Always check the current legislation on the markets in scope.

Natural flavoring:
Natural flavoring is when a flavoring is derived from natural sources and processed by using physical, enzymatic, or microbiological processes. A declaration, e.g., may read: "Natural flavoring," or "Natural vanilla flavoring" (if the flavoring is derived exclusively from vanilla).

Flavoring:
When the flavoring is either synthetic or does not meet the criteria to be labeled as natural, the declaration will be: "Flavoring," or "Artificial flavoring" (though typically just "flavoring" is used).

Smoke flavoring:
When the flavoring is derived from purified condensed smoke, the declaration will be: "Smoke flavoring," or "Smoke flavor."

Masking flavors:
Masking flavors are ingredients added to a food or beverage product to reduce or eliminate unwanted tastes, such as bitterness, astringency, metallic notes, or overly intense flavors. These are commonly used in formulations with functional ingredients like plant-based proteins, vitamins, minerals, caffeine, or artificial sweeteners, which can have off-putting taste characteristics.

Specific natural flavoring source:
When the flavoring is derived from a specific natural source, the declaration will be: "Natural strawberry flavoring," if at least 95% of the flavoring component is derived from strawberries. If it is less than 95%, but the characteristic flavor is from strawberries, it must be labeled as "Natural strawberry flavoring with other natural flavorings."

Natural Products

Many natural products have complex flavor profiles, often so complex that a flavor wheel[7] is used to map out the different aromatic directions. Fermented products in particular develop a wide range of volatile compounds, contributing to their depth and complexity. However, not all volatile compounds are desirable. Some create lingering aromatics that result in unwanted off-flavors or even spoilage. Some well-known fermented products include cheese, yogurt, miso, beer, and wine.

Wine

Wine is actually an area where identification of flavors is taken very seriously. A *sommelier* is a trained and knowledgeable wine professional who specializes in all aspects of food and wine pairing. They traditionally work as waiters and wine experts in fine dining restaurants.

To become a sommelier, one has to go through an extensive training program. Their sensory training includes blind tastings to refine the ability to recognize grape varieties, regions, and winemaking styles. They train their palate by tasting a wide range of wines, analyzing them, and developing an olfactory memory by identifying common flavors in wines, such as fruits, spices, and earthy notes. Their expertise extends beyond wine and often includes spirits and beer.

Beer

Beer is a great example to highlight the multitude of flavors that can be developed by using craftsmanship and brewing techniques on a fairly few and simple ingredients. Some of the flavors that can be identified in beer are the following:

*   Malty flavors, like toasted bread and biscuits.
*   Sweet notes, like honey, caramelized sugar, vanilla, or chocolate.
*   Fresh, tangy and aromatic notes, like lemon zest, grapefruit or crisp apple, berries or tropical fruit.
*   Floral and herbal notes, along with warm spicy notes like cloves, cinnamon, and nutmeg.

As illustrated, sweetness and sourness can also be perceived through flavor, simply because we associate the flavor of certain ingredients with their inherent basic taste. That also goes for salty, bitter, and umami.

SIGHT

The visual appearance of F&B products plays a crucial role in shaping consumer expectations and in influencing perceptions of freshness, quality, and desirability. Seeing the packaging gives consumers their first impression of the product. The appearance of the product packaging provides a lot of visual cues: by the text, images, and colors. There is often a gap between what is communicated on the packaging and the experience when opening a food or beverage product. Packaging design, imagery, and branding set expectations, but the real appearance of the product will either confirm or contradict those expectations, directly influencing consumer satisfaction. Some products do, however, have transparent sections of the packaging, allowing the consumer to see and assess parts of, or the entire product, before purchasing it.

During consumption, the visual appearance of a product, meal, or beverage is even more important. The way something looks before you eat or drink it has a strong impact on the taste experience. In some cases, the appearance of food and beverages seems to matter more than the actual taste. Think about all the pictures people post of their food on social media channels. There is even a term for it, called *food porn!* The term gained mainstream popularity in the 2000s with the rise of food blogs, social media, and platforms like Instagram, where visually indulgent food photography became a cultural phenomenon.

Designing retail F&B products to look as delicious as possible is always an aspiration in innovation, but also a struggle. It is hard to compete with the visual overload from the media, but also from within the industry. The images on the packaging and in commercials often oversell the actual visual appearance of the products, leading to disappointment. Most people know not to expect a homemade lasagna to magically appear when opening a frozen ready meal, but I suppose they always hold onto the hope that it might!

Touch

The tactile sensations of a product play a crucial role in shaping the overall sensory experience. Texture, consistency, and their associated mouthfeel influence how a product is perceived, beyond taste and flavor. A product can, for example, feel smooth, creamy, watery, crunchy, dense, airy, chewy, etc. All sensations affect its appeal and enjoyment. There are also certain expectations for the tactile sensations of specific products. You don't want your cracker to be chewy, your cupcake to be dense, or your smoothie to be watery. Right?

Some of the mouthfeel sensations include the following:

- **Smooth** textures like creamy yogurt or velvety mousse.
- **Gritty** sensations, found in foods like polenta.
- **Grainy** textures such as those in pears. They have noticeable small particles, but they aren't as coarse as gritty textures.
- **Crispy** foods such as potato chips.
- **Crunchy** items like granola or nuts, that provide a firmer bite.
- **Chewy** foods such as caramel or mochi.
- **Gummy** textures like gummy bears, that have an elastic, bouncy sensation.
- **Gel** textures, such as panna cotta or jelly.
- **Tender** foods such as ripe avocados, are soft and easy to bite through.
- **Juicy** foods like watermelon or steak, that release liquid upon biting.
- **Watery** foods such as skimmed milk or jelly.
- **Fibrous** textures like celery or mango.

- **Powdery** foods like cocoa powder or powdered sugar, that are fine and dry, often coat the mouth.
- **Sticky** foods such as peanut butter or honey.
- **Oily** foods like butter or fried items.
- **Foamy** foods such as whipped cream or mousse, that are airy and dissolve quickly.
- **Bubbly** textures found in sparkling water or beer.
- **Astringent** foods like unripe bananas or tea, that create a dry sensation.

Not all tactile evaluation of products happens in the mouth, and it is also very relevant to consider how the products feel in the hands. For example, crumbly foods like feta cheese or biscuits break into small pieces easily, and it is part of their product identity. Bread should be firm enough to be cut into slices, a marshmallow will reshape after the release of pressure, and so on.

In addition, I would also like to point out the tactile importance of fat, which enhances the taste experience immensely. The consumption temperature and the "heat" created by chili and other ingredients with similar properties.

Fat as a Tactile Sensation

As a fundamental tactile sensation, fat contributes to the richness, smoothness and overall enjoyment of a product, affecting both how it feels in the mouth and how flavors are delivered. Creamy and smooth textures are often the result of the fat content in foods like butter, ice cream, or high-fat yogurts. This creates a luxurious, indulgent sensation that is often associated with premium or satisfying foods. The melting or coating properties of fat affect how food breaks down in the mouth. Chocolate, for example, melts at body temperature, providing a smooth, gradual release of flavors. Whether providing creaminess, richness, or a lasting flavor experience, fat plays a central role in sensory enjoyment. In understanding how fat interacts with temperature, flavors, and textures allows for strategic product design, ensuring the optimal sensory impact for consumers, and bearing in mind that a high fat content is both associated with comfort and indulgence, but for some consumers, with unhealthy products.

Temperature as a Tactile Sensation

Some products are stored in ambient conditions, but should be served cold and some are stored cold, but should be served hot. It can get complicated ... One of the big challenges for a "drinking yoghurt", for example, is that it is hard to bring it "on the go", as the temperature will increase quite fast and make the taste experience less enjoyable.

Temperature influences both the physical sensation of a product and how its flavors and textures are perceived. The serving temperature of food and beverages, whether warm or cold, triggers specific sensory reactions, impacting how consumers experience the product. Understanding how temperature influences food and beverage sensations is also important in designing the product and ensuring that consumers experience the intended sensory impact.

Spiciness (heat) as a Tactile Sensation

The sensation of heat from chili peppers, mustard, horseradish, and similar ingredients is a unique tactile experience that is distinct from temperature-based heat. This type of spiciness is a chemical irritation, detected by pain receptors. It triggers a burning, tingling, or warming effect that varies in intensity and duration, influencing how a food or beverage is perceived, both physically and emotionally. Spiciness is perceived as a pleasurable or even addictive experience for many consumers. There is also a deep emotional connection to certain food cultures worldwide, where heat is deeply ingrained in various cuisines from the spicy Vindaloo curry from India, the ruthless green papaya salad from Thailand to a Salsa de Chile Habanero from Mexico.

Hot sauces, in general, are widely used globally. The most famous is the Tabasco Sauce.[8] Tabasco Sauce originates from Avery Island, Louisiana, USA. It was created in 1868 by Edmund McIlhenny, who used tabasco peppers, salt from Avery Island, and vinegar to craft the hot sauce. Yet another example of the wonders of creating delicious aromatic products by fermentation. In this case, lactic acid fermentation.

HEAR

Sound plays a surprisingly important role in the way consumers perceive and experience food and beverages. The auditory cues, both before and during consumption, can significantly shape expectations, enhance the enjoyment, and even affect the perceived freshness and quality. The first sound interaction often comes from the packaging. The snap of a soda can opening, the crunch of a chip bag, or the pop of a vacuum-sealed jar creates an immediate sensory expectation. These sounds serve as confirmation signals, reassuring consumers that the product is fresh and properly sealed. Some brands even engineer packaging sounds deliberately to enhance the sensory experience.

During consumption, the texture and structure of food influence the sounds it produces, which add an extra layer to how we experience flavor. A crispy bite, a crunchy snack, or the sound of a sparkling beverage being poured contribute to the overall sensory experience. Just like with visuals, there is often a gap between expectation and reality when it comes to sound. Advertisements exaggerate the crispy crunch of a snack, or the intense fizz of a drink, setting an expectation that might not fully match the real-life experience. While consumers generally recognize these audio enhancements, they still respond positively to them, reinforcing their desire for multisensory satisfaction.

Sound is an underrated but essential part of food and beverage innovation. Whether it's the sound of a package opening, the texture-driven crunch, the auditory experience can significantly influence how much consumers enjoy and even remember a product.

SENSORY PANEL TESTING

Measuring, understanding, and controlling organoleptic properties are essential for product design to meet consumer preferences and quality assurance. In other words, the tests and results are used not only in the innovation process, but also in the daily operation of an F&B producing company, to ensure that the product also tastes and looks as expected when released from production.

Organoleptic evaluations are, in essence, subjective and rely on human perception. That said, product characteristics can also be measured by using different instrumental and physicochemical methods. Sensory panels are typically composed of highly skilled professional tasters, hired to evaluate products. Some of them are even referred to as super tasters. They are individuals who have an enhanced sensitivity to certain taste compounds. This not only makes them more sensitive to the basic tastes like bitter, sweet, salty, sour, and umami, but also to the flavors.

Many different tests can be conducted by a sensory panel. The panelists are trained in identifying the taste experience and describing it. They train a common vocabulary by tasting and agreeing on what to call the taste, flavor, texture, and so on. Using sensory panels is a means to have a professionalized and hopefully unbiased evaluation of the taste experience. Some of the large F&B companies have their own panel of trained sensory evaluators. If not, it is possible to hire companies that specialize in sensory evaluation to assess the products. There are many methods used in sensory science, and selecting the best method will be very project and product-dependent.

QUANTITATIVE DESCRIPTIVE ANALYSIS (QDA)

Trained panelists evaluate products and describe their sensory attributes using standardized terminology and scoring scales. A QDA[9] is a method used to describe the sensory characteristics of a product in a detailed and objective way. A trained sensory panel evaluates products and rates how intense different attributes are. It could be sweetness, creaminess, or acidity – using a numerical scale. The goal is not to say whether a product is good or bad, but simply to measure how much of each sensory quality it has. There are different ways to visualize and interpret the results of a QDA. One common way is the spiderweb chart, which shows the intensity of each attribute per product in a clear, individual profile.

SPIDERWEB

The *spiderweb* (also known as a *spider plot* or a *radar chart*) is one of the known ways to display sensory attributes of a product. It is a way to visualize and compare different sensory characteristics or attributes of a product in a comprehensive and easily interpretable manner. Spiderwebs are particularly useful in sensory evaluation for comparing products and communicating the results in a clear and visually appealing way.

As illustrated in Figure 4.1, the key characteristics of a spiderweb in sensory evaluation include the following:

Axis: Each axis represents a different sensory attribute relevant to the product.

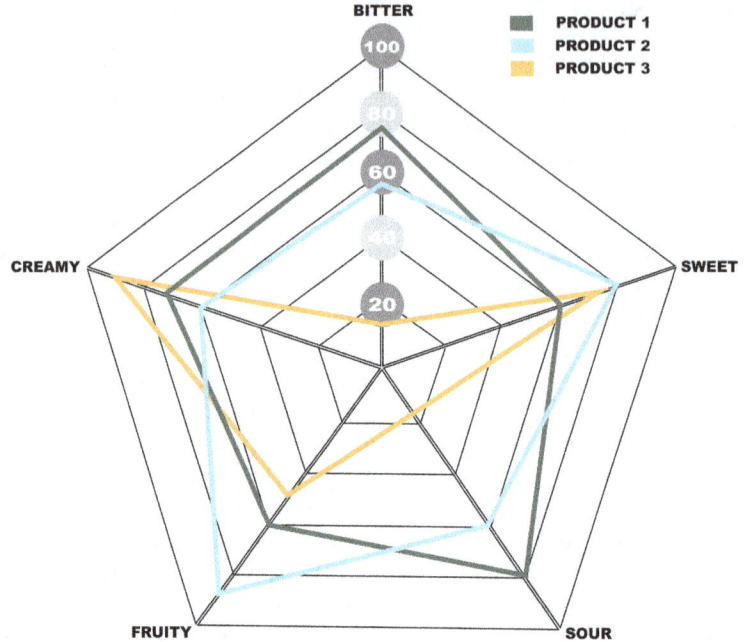

FIGURE 4.1 Spiderweb Chart of Sensory Attributes.

Scale: Values are usually plotted on a standardized scale (e.g., 0 to 100) along each axis.

Shape: The shape of the web gives a visual representation of the product's sensory profile, showing similarities and differences in the various attributes.

Comparative Analysis: Multiple products or samples can be plotted on the same spiderweb for comparison, with different colors or line styles representing each sample.

PRINCIPAL COMPONENT ANALYSIS (PCA)

Another method is a PCA,[10] which is a more advanced, statistical approach that summarizes the sensory data across all products and attributes. PCA creates a map that shows how products relate to one another based on their sensory profiles. This helps to identify patterns, similarities, or clusters across products. Imagine placing all the products on a two-dimensional map, where products that are close together taste similar and products that are far apart have very different sensory profiles.

The map also shows which sensory attributes (like sweetness or creaminess) are driving the differences between the products.

CONSUMER TESTING

To understand the consumers' preferences, consumers can also test and provide sensory feedback. One of the very prevalent methods is the drivers of liking.[11]

DRIVERS OF LIKING

Drivers of liking refer to the specific attributes of a product that connect with the consumer's preference, and hence, enjoyment. These can be measured through various sensory evaluation methods, capturing how different sensory attributes affect the overall liking. Samples of prototypes, or existing products, can be served to consumers in smaller focus groups, or larger groups, alongside competitor products. The products will be served blindly, and a questionnaire will ask the participants about their taste experience. The focus will be on rating the sensory attributes, as well as the perception and overall liking. Depending on the product category, focus will be on the areas that make sense.

As an example: If you are testing natural yogurt prototypes, together with some competitor products, it would make sense to ask about the perception and liking of the following:

- The flavor profile
- The creaminess
- The sweetness level
- The acidity level
- The freshness
- The thickness

You can also use this opportunity to ask about their purchase intent. How likely they would re-buy the specific yogurts, based on the taste experience. Focus groups normally don't have a lot of respondents, and if you feel like having a larger group to base your decisions on, a central location test (CLT) can be done. It can also, in some cases, be possible to ship the samples to people's homes. A home use test (HUT). The consumers in a home use test can then evaluate the products and provide their feedback in an online survey.

More about CLTs and HUTs will be discussed at a later stage.

N.B. It is very important to understand the importance of the field of sensory and consumer science to enable F&B innovation. This is an area that is sometimes given lower priority in businesses and this can lead to designing the product experience based on personal preference, and simply hoping it aligns with the preferences of the target audience. Which is not recommended …

The next topic and chapter focus on the food and beverage industry – exploring the wide range of company sizes and types within it. It examines how these companies depend on and support one another.

NOTES

1   *Data Science for Sensory and Consumer Scientists*, Thierry Worch; Julien Delarue; Vanessa Rios De Souza & John Ennis, Chapman & Hall/CRC, London, 2023.

2 *Fornemmelse for Smag*, Ole G. Mouritsen & Klavs Styrbæk, Nyt Nordisk Forlag, Copenhagen, 2015. Page 1–8.

3 *The Biological Basis of Food Perception and Acceptance. Food Quality and Preference*, L. M. Bartoshuk, Elsevier, The Netherlands, 1993.

4 *Koji Alchemy: Rediscovering the Magic of Mold-Based Fermentation*, Jeremy Umansky and Rich Shih, Chelsea Green Publishing, Chelsea, Vermont, 2020. Page 267–270.

5 www.jpo.go.jp/e/introduction/rekishi/10hatsumeika/kikunae_ikeda.html

6 https://food.ec.europa.eu/food-safety/food-improvement-agents/flavourings/eu-rules_en

7 Although commonly called a "flavor wheel," some flavor wheels integrate descriptors beyond flavor, covering taste elements like acidity, sweetness, and bitterness, as well as texture and mouthfeel.

8 www.tabasco.com/tabasco-history/

9 *Sensory and Consumer Research in Food Product Design and Development*, Howard R. Moskowitz; Jacqueline H. Beckley & Anna V. A. Resurreccion, Wiley-Blackwell, Hoboken, New Jersey, 2006.

10 *Sensory and Consumer Research,* Moskowitz et al., 2006.

11 *Sensory and Consumer Research,* Moskowitz et al., 2006.

5 Food and Beverage Industry

TYPES OF COMPANIES

The food and beverage industry can be defined in many ways. The broadest definition would be all companies that capitalize on food or beverages in some shape or form.

The following discusses some examples of industries and how they think and operate:

CAFE/RESTAURANT

If you have a small cafe or restaurant, you are a part of the food and beverage industry. The menu they offer their guests can be designed in many ways. Many owners and chefs create the dishes they want to put on the menu all the time. They also decide on what beverages, wine, and beer, go along with the food. These creative choices are many times made without a deeper understanding of the potential guests but are based on personal preference and a mindset of "that's how we normally do things".

Imagine how it can be if more effort were put into understanding the target audience and if new dishes were tested and adjusted according to the feedback. There lies a big potential in working more structured with design thinking in this part of the industry. However, it is **not** the focus in this book. The target audience of this book is the traditional food and beverage industry, producing and selling their products in retail, but also the adjacent industries that support the F&B innovation.

The industry is basically divided into two types of companies, B2C (business to consumer), and B2B (business to business).

In B2C companies, i.e., companies that produce the final product for the consumers, typically sold in retail, it is important to understand the consumers in the markets they operate in or aspire to enter. It is also important to understand the competitive market dynamics and distribution channels, as well as how to communicate what the products are about and why consumers should buy them.

B2B are companies that produce ingredients or services for B2C companies. In B2B companies, it is important to understand the businesses of their customers. Also important is to understand their needs, priorities, and struggles, including how they make decisions and influence procurement. They also need to understand the competitive landscape and how to communicate what their products or services are about,

DOI: 10.1201/9781003619352-6

and why they can help their customers to solve problems, improve performance, or gain a competitive edge.

Both types of companies come in many shapes and sizes. There are small startups to large global companies with factories and activities in many countries. Some companies also have a mix of a B2C and B2B business. All-sized companies can be active in several markets, but the bigger the companies become, the more global they normally become. Let's have a look at some of the different types of companies.

STARTUPS

An F&B-startup typically has between 1 and 12 employees, and it can take many different shapes and forms. The main characteristic is that they are driven by a passionate individual or individuals. The purpose and ambition can, however, be very different. Some startups are driven by a single product idea, created by the founder. That idea might have come from the founder's ability to create something delicious based on a food trend, an observation of shifting consumer behavior, or maybe a desire to promote more sustainable eating.

Startups may still be exploring their business model. They could be starting their journey with a launch in retail, but later in the process, conclude that their profitability would be better in B2B. As an example of this, I would like to share a story about a startup that I first met in 2021: the company Reduced.[1]

In the summer of 2024, I met Emil for an interview about their journey and how they work with innovation. Earlier, in 2020, Emil Munck de Voss and William Lauf Olsen founded the company, *Reduced*, based in Copenhagen. A successful startup that in 2024 secured €6M in further funding from a group of investors. They primarily work with food side-streams, also known as *waste*, from the food industry, but they also work with ingredients that normally are not used commercially, such as beach crabs. They turn these side-streams and other raw materials into natural food ingredients through fermentation. They have several different products in their portfolio. The products are based on all kinds of different sources of meat, fish, and plants.

One example of a product is a mushroom-umami sauce. They buy the waste from the production of mushrooms and transform it via fermentation into a flavorful liquid with a unique umami taste similar to a soy sauce. They started with small bottles for retail, but lately, they have moved the business more into the B2B part of the industry.

They currently target two segments:

1. Food service distributors that sell to restaurants and canteens. In this sector, their shellfish broth, made from beach crabs, has been a significant success.
2. Food manufacturers that use the natural flavor solutions for sauces, soups, and ready-made meals.

They are focused on the major trend of reducing food waste, which has seen significant growth in recent years. Addressing food waste not only prevents the loss of valuable nutrients but also optimizes the production processes by utilizing side-streams that would otherwise go unused, and thus ultimately reduces the carbon footprint.

Another key unique selling point is the perceived naturalness of the products. The natural fermentation process they apply to the ingredients enables them to fit with the industry's desire to communicate "clean label" products.

Being a new player in the B2B food industry is not easy, by any means. The competition from large ingredient companies with long-standing customer relationships is fierce. They sell solutions that provide the same functionality, and at a cheaper price. Attractive price points are essential to success in the B2B industry. Reduced's focus is to emphasize the appealing story around reducing food waste and sustainability, delivered in the form of delicious, natural taste solutions. The more successful they become, the more efficient production processes they will be able to make and hence be more price competitive. Their advantage over large ingredient companies lies in their agility to tailor products to meet their customers' specific needs and to produce smaller batches.

Meeting and talking with entrepreneurs like Emil is very inspiring. The world of startups and their struggles to succeed is fascinating and forms the entire foundation of the food and beverage industry.

Regardless of the size of the company, it is the passion and people behind it that drive successful innovations. Capabilities change according to size, but there is always a need for all sizes and types. Small and agile companies have their advantages. Medium-sized ones offer more muscle, certifications, and a degree of agility while large companies benefit from scale and competitive prices, but often at the expense of flexibility.

SMALL COMPANIES

A small company could of course also be a startup – at some point, most new companies are. The small company would have between 1 – 50 employees. Some of the small companies have their own production lines, while others outsource all or part of their production. In many cases, there's a mix of both approaches. Innovation is done with a *hands-on* approach, typically in collaboration between marketing, sales, procurement, and a product developer.

Small B2C companies are normally more agile and have more speed to market with the new products they wish to launch. They most likely only have one brand and limited resources for marketing. They are very dependent on good relations with retail and network.

Small B2B companies are probably also very specialized in the product solutions they provide to their business customers. Often, they are driven by briefs for their customers, but they could also have a focus on growing the business through innovation.

MEDIUM-SIZED COMPANIES

The medium-sized company would have between 50 – 500 employees. They normally have their own production lines but may outsource some of the production. Innovation is most likely conducted in a more structured setup, still typically a collaboration between marketing, sales, the product development department, and other essential functions in the company. Some might even have an R&D department.

Medium B2C companies are normally more established with sales distribution: they have customers who negotiate and renew contracts on core products each year. One or more brands are normal. Each year innovation pipelines are built and executed.

Medium B2B companies have more variety in the product solutions they provide their business customers. They are also often driven by briefs for their customers, but will most likely, additionally, have a proactive focus on innovation and resources allocated to do so.

Large Companies

The large company would have between 500 – 20,000 employees. They have their own production lines. They could have multiple factories with several different production lines and they may also still outsource some of the production. In many cases, innovation will have its own R&D department, including product development, consumer insights, and sensory and consumer science teams. Innovation is a collaboration between marketing, sales, other essential departments in the company, and the entire innovation organization.

Large B2C companies are very established with sales distribution and have customers who negotiate and renew contracts on core products each year. They have several brands in multiple markets. Each year, innovation pipelines are built and executed.

Large B2B companies, on the other hand, have a huge number of very specialized ingredients and product solutions. They are often driven by briefs for their customers and will have a focus on proactively growing the business through new product innovation, and other services, for which they have many resources to do so.

SUPPORTING COMPANIES FOR F&B INNOVATION

Supporting companies don't launch and sell food products, but they play a vital role in the food and beverage industry by providing capabilities that go beyond the expertise of the employees and available equipment. Depending on the size and type of the F&B company, the internal capabilities and resources vary. Many times, F&B companies don't have the luxury to invest heavily in hiring all the skilled people they would like to. However, I will recommend F&B companies to prioritize their capabilities and resources to handle and take ownership of their own innovation processes, to the extent that it is possible.

Medium- or large-sized companies should prioritize investing in internal expertise, while small companies or startups need to have a DIY mindset in many aspects. Regardless of how well an F&B company is structured, all in the industry rely to some extent on supporting companies to inform or enhance their innovation processes. It is just important to strike the right balance between leveraging the internal company intelligence and seeking support from external expertise.

In the chapters on nutrition science and sensory and consumer science, we have already gone through some of the key areas of knowledge you need to have in the F&B industry, but there are many more to come. There will be more details about the knowledge you need to have, to take informed innovation decisions in the chapters to come, but as a short teaser, here are some key areas that external companies typically can help you with.

INNOVATION PROCESS CONSULTANCY COMPANIES

These companies specialize in designing and facilitating innovation processes. They help structure ideation, run creative workshops, and support the development of new concepts. Their strength lies in guiding teams through the innovation journey with tailored tools, proven methods as design thinking, but also providing an outside perspective. Along the way, they often challenge assumptions and ask just the right questions to spark new thinking and energize a team.

CONSUMER RESEARCH COMPANIES

Consumer research companies focus on understanding consumer needs, preferences, and behaviors. They gather insights through qualitative and quantitative methods to inform product design, branding, and communication strategies.

DATA COMPANIES

These companies provide data that can support innovation decisions – from competitor sales performance to shifts in consumer purchasing behavior. They help companies identify market gaps, track product trends, and assess category dynamics. Services often include access to dashboards, analytics tools, and custom-made reports based on retail data.

TREND AND FUTURE FORESIGHT COMPANIES

These companies help identify and interpret both consumer behavioral trends and food and beverage trends. They can provide early indicators of change and possible future scenarios. Their insights support innovation and strategy work by helping businesses anticipate change and explore future opportunity spaces.

BRAND AND MARKETING COMPANIES

These companies focus on shaping how a brand is perceived. They help develop brand platforms, visual identities, and marketing campaigns. Their strength lies in translating strategic goals and consumer insights into creative and compelling communication.

Packaging Design Companies

These companies specialize in the creative and functional aspects of food and beverage packaging. They design packaging that not only catches the consumer's eye on the shelf, but also communicates key product expectations, such as taste, quality, or health benefits – before the product is even tried. Their work also plays a critical role in expressing brand identity and positioning.

Sensory Testing Companies

These companies help to evaluate how products are perceived through the senses: especially taste, smell, and texture. They provide structured testing with trained panels or consumers to help refine product recipes and ensure consistency and quality, in achieving the best possible taste profiles.

Full-service Consultancy Companies

The full-service companies can mostly provide everything from consumer research to shaping the company and brand strategy. They cover more or less all the areas from above. What they can't do themselves, they outsource to close collaboration partners. They are the most convenient choice!

Digital Tools and External Support

Throughout the chapters in the book, I will touch on various digital online tools and software, but I will refrain from being specific about names and companies. The functionality will be described in broad terms, as well as how the tools might help you in the innovation process. It is mainly to avoid referring to a tool or company that, in the long run, will change names, be surpassed by a competitor, or suddenly become completely obsolete. Technology moves very fast, and my advice is to look for tools that you find useful, have good reviews, and subscribe to them when they meet your needs.

The same goes for hiring supporting companies. They will also be changing their shape and form in the same way. The rapidly advancing technologies will provide them with new "products" to offer, and these can and should be used to the extent that it is viable for your project. There are many companies out there, and it is hard to navigate. The quality of the offerings comes in many levels. Prices can vary from affordable to sky high! The assessment of the resources to provide your project with the building blocks to create great innovation, will depend on many factors in the interaction between the nature of your company and the nature of the potential market opportunity. In all cases, make a budget and prioritize the external activities according to the universal rule of "Nice to have" and "Need to have."

PRIVATE LABEL – RETAILERS

Retailers often work closely with manufacturers and agencies to customize their private label assortment. Depending on the specific country and retailer, private label is prevalent across markets, and the business model is growing. The retailers' own brands come in many shapes and forms. Most big retailers have a price ladder approach: a budget range, a mid-tier range, and a premium range. This price ladder approach allows retailers to cater to different customer segments, offering products that vary in quality and price to meet diverse consumer needs and purchasing power. In terms of the product design, the retailers have essential data available about what to launch, and the focus on understanding consumer behavior. Based on their sales data, they can track the success of all the products they sell. This enables them to assess the potential of a product and what to launch. Balancing this situation can be challenging, especially if a top-selling product comes from a brand that closely collaborates with the retailer. While B2C companies often have their own branded products, they also frequently provide private label options to retailers.

PLANT-BASED DRINK EXAMPLE

Imagine that a medium-sized B2C company has launched a new product under its own brand. They want to sell it to retailers, but at the point in time, the product type falls into a somewhat niche category. It could be a plant-based drink in early 2016. The retailer acknowledges that an emerging trend of more sustainable food solutions is real and shows potential to grow. They offer the company shelf space on their "spot deal" shelf. This shelf features time-limited offers used by the retailer to test new products with low risk for them, and for the companies behind – the chance to "get in" and prove that the product is attractive to consumers.

In this case, the consumers buy the new plant-based drink in good quantities, and the retailer offers the company a permanent listing if they develop a range of 3 – 4 variants of the plant-based drink. The company was very happy with the successful launch and had started the development of a range. In 2017, they were in all the stores of the retailer and the sales figures exceeded the volume expectations. During 2018/19, sales became significant enough to motivate the retailers to launch a plant-based drink range under their own mid-tier brand. Unfortunately, this resulted in the delisting of the original company brand. They were asked by the retailer if they wanted to supply the range in the mid-tier private label brand instead, and they accepted this. The price point is lower, and hence the profit for the company will be lower.

The tension here is obvious, but the company will risk losing the current volume and perhaps future business if the offer is refused. This is normal business in the F&B industry. If the original brand is strong, a drop in sales can happen. If the product quality is kept, the price is lowered, and the private label brand has a similar appearance, volumes are likely to increase.

CUSTOMER VS. CONSUMER

Customer and *consumer* are two terms used in literature and within the industry to describe the end user of the products. *Customer*, however, is also used to describe the companies that buy the food and beverage products from the producing companies. It can be quite confusing! In this book, the term *consumer* will always refer to the individuals who ultimately buy and use the product. It may sound simple, but there are important nuances to keep in mind.

DRINKING YOGURT EXAMPLE

If you buy a drinking yogurt for yourself, you are both the buyer and the consumer. But if you buy it for your child, you are only the buyer, purchasing on behalf of someone else, who will consume the product. What you, as a parent, find important might not align with what your child finds important!

The Parent (Buyer) Perspective

As a parent, when buying a drinking yogurt for your child, it is likely that you focus on the following:

- **Health and Nutrition** – Is it good for my child? Does it contain essential vitamins, calcium, and probiotics, or is it reduced in sugar?
- **Convenience** – Is it easy for my child to drink it on the go, take it to school, or finish it quickly?
- **Value for Money** – Am I getting a good deal in terms of price, portion size, and benefits?
- **Trust and Safety** – Is it from a reliable brand with ingredients I can trust?
- **Picky Eating Considerations** – Will my child actually like this flavor, or will it go to waste?

The Child (Consumer) Perspective

From a child's perspective, consuming the drinking yogurt may have different priorities:

- **Taste and Flavor** – Do I like how it tastes? Is it sweet, fruity, or chocolatey?
- **Fun and Experience** – Does it have a cool design, a fun straw, or interactive packaging (e.g., cartoon characters)?
- **Peer Influence** – Do my friends drink this? Will it make me feel part of the group at school?
- **Instant Gratification** – Does it feel like a treat or something enjoyable rather than just "healthy?"

Balancing these different functional and emotional needs and motivations can be difficult for the parent, but even more so for the company designing it.

WELCOME DRINK EXAMPLE

In another situation, imagine that you are hosting some friends for a cozy get-together and buying a variety of snacks to serve alongside a welcome drink. Instead of choosing only what you personally enjoy, you consider a broader selection to accommodate the different taste preferences of your friends. You might pick options that cater to dietary restrictions of your friends, popular or new flavors, or items that complement the drinks being served. Even though you are making the purchasing decision, your guests are the ones consuming the snacks, and your buying motivation is based on hospitality and variety, rather than your personal preference.

These and similar examples are still within my definition of consumers. They are still the end user, in the sense that they make buying choices of products to serve their friends and family. It is the level of the relationship between the buyer of a product to the end consumer of the product that makes the difference. You become a customer when you buy the products with a more professional aim.

BAR OWNER EXAMPLE

Let's revisit the example of serving guests drinks and snacks. If you own a bar and are buying the assortment of snacks you want to sell to your guests, then you are a customer. You are not necessarily a customer of the producing company, but the retailer or wholesaler, having the products in their range. For the producing company, the customer is the retailer or the wholesaler. Sometimes, there are several links in the value chain, before the end user (the consumer) eats or drinks the product.

In design thinking the consumer is at the center of innovation, but the need to innovate against the needs of your customer is also very important.

In this book, the customer is the distributor or the retailer buying the F&B products in bulk. Customers also have wishes and requirements for the products they buy and sell to the end consumer.

RETAILER EXAMPLE

Retailers have many things to consider when they build their product assortment. Firstly, they have the constraints of the shelf space in the physical shops. All the different categories have a certain amount of space and requirements for the size and shape of the product packaging. The packaging may need to be stackable and possibly to be placed in a visually appealing way. Retailers, of course, also have a high focus on their target audience and seek to have an assortment that aligns with their preferences and purchasing behaviors. They try to balance staple items with unique or niche items to attract diverse customer segments and keep the shopping experience interesting. In the case of private label products, they might control these wishes themselves, but in many cases, the procurement department will have meetings with sales employees from the producing companies and brands, where they will ask for specific new types of products.

As an example of how this can look , I can share a nice story from one of the big retail players[2] in Denmark. In 2006, they launched a new own label range of organic products. By doing that, they emphasized their identity as a responsible company towards health for both people and the planet. The kind of food you would love to buy for your children, and be seen buying by your neighbors. This meant unique possibilities for suppliers of organic products, and I am sure that the sales force from these companies returned to their offices with the message to invest in capabilities to produce with the organic certification. It was a successful initiative, and the retailers increased their organic sales year after year. In 2015, they set a goal to double the sales of organic products from 2015 to 2020. A very ambitious goal.

This goal opened more shelf space in their shops than before, and they dedicated entire aisles of shelves for the organic food products. One of the well-known organic brands in Denmark got the opportunity to have those shelves, if they could create a suitable assortment quickly. They could and they did! The combination of their own label and the well-known brand assortment generated sales over the next five years to fulfill the ambitious goal. The retailer had the insight to put "organic" into their strategy and make the investment. The companies around them, ready to deliver organic products, had a huge advantage. This strategy was effective in Denmark but may not have succeeded in other markets where consumers didn't have a similar unmet demand for organic products. The Danish government has also played an important role in supporting the organic sector through robust certification and labeling systems, which have significantly boosted consumer confidence in organic products.

The definition of consumer and customer is now in place, as well as all the introductions to what the food and beverage industry is about. Let's summarise the main components to take into account:

- The products and how to store them.
- The nutritional aspects of consuming them.
- The joys of the taste and product experiences.
- The industry and some of the breakthrough innovations so far.

Let's enter the core topic of this book:

How to apply the design thinking methodology for F&B innovation.

NOTES

1   https://reduced.dk/
2   https://info.coop.dk/ansvarlighed/okologi/

Section II

Understanding Design Thinking

6 Design Thinking

DESIGN THINKING FOR F&B INNOVATION

The heart of design thinking is understanding human behavior through empathy and translating people's needs, motivations, and barriers into solutions that add value to their lives. As mentioned, design thinking has increasingly become a popular innovation tool in a range of industries. It has especially excelled in the IT industry, where user experience can easily be iterated through fast prototyping and consumer testing. IT companies even launch demos on the market and get feedback and data from their users.

The IT industry is pretty unique, as they can launch products and, in real time, adjust their product to have better functionality, safety, and visual identity, several times a year! How many times haven't you tried to get a "new update available" notification on your phone, laptop, or even in your new car?

Other industries, like the toy industry, are also really embracing the methodology in understanding how children play, providing them with the empathy needed to innovate new engaging toys. LEGO from Denmark, as an example, is well known for using design thinking in their innovation. In the IT software industry, it is very common to identify a consumer tension before even considering product ideas. IT software is very moldable, and you have few limitations in terms of creating a product that could solve a specific problem. It all depends on the team's imagination and coding skills.

Prototypes can be created by writing lines of code into a visual interface and testing them online with consumers. The same day, you can have feedback and adapt the prototype accordingly, and test again. It is a fast and flexible process.

FOOD AND BEVERAGE INNOVATION IS A DIFFERENT BEAST!

In the food and beverage industry, prototypes and how to test them with consumers are two of the most important aspects of the design thinking methodology, which is significantly more complex than in the IT software industry.

DOI: 10.1201/9781003619352-8
49

Prototypes can be made in various ways, but at some point you will need to create a physical prototype, which is:

- Safe to consume.
- Visually appealing and tasty.
- In a relevant packaging that conveys the desired expectations.
- Feasible to replicate at the scale needed.
- Viable to produce in terms of the cost of production, and to be sold at a realistic sales price.

There are key attitude and behavioral factors to bear in mind when receiving feedback from showing prototypes to consumers:

Consuming food, drinking water, and other beverages is a common practice for everyone. It's something everyone experiences every day and has an opinion about!

It is a skill that we train every day, and it involves a complex interplay of sensory, emotional, cognitive, and physiological processes, with various regions of the brain working together to assess and interpret the experience.

- It is a multi-sensory experience consisting of smell, taste, texture, and visual appearance, as we touched upon in the sensory and consumer science chapter. Consumer preferences are very subjective, and also culture- and occasion-dependent.
- It is a functional experience that provides the body with the feeling of satiety and the nutrients to keep it alive but also to adjust the health and cognitive performance.
- It is an experience that interlinks and activates memories of all previous sensory experiences, however, good or bad they might be. This also means that a certain taste, texture, or smell can trigger unconscious reactions and attitudes towards it.
- It can be a social experience in some situations and in others, a purely personal experience.
- It is used to convey your personal identity and your ethnic heritage.
- The amount and quality that people consume are dependent on the socio-economic situation of the individual.

Furthermore, and essential to deeply understand and acknowledge, is that food and beverage innovation is not merely about getting a great idea and creating something edible or tasty. It's an intricate process that involves culinary skills, food science, sensory and consumer science, all kinds of market and consumer research, acquired consumer empathy, production process engineering capabilities, commercial flair, corporate ingenuity, packaging design capabilities, and great marketing. This is just to name a few areas, areas of expertise to be mixed with great creativity! In fact, it is actually the creative process based on this diverse foundation of different skills and areas that can generate innovation.

You might question whether design thinking is useful in both radical and incremental innovation? For sure, one can use some parts of the design thinking methodology to make incremental innovation better. Many of the processes and tools are useful. One only has to have a clear starting point and should be conscious about what type of innovation the project is about. One should be mindful not to use unnecessary resources, if it is just a line extension within a category of well-established products. It is a well-known mistake to start a project with a specific product in mind, pretending to look more broadly and explorative into the category, or even adjacent categories. It is a waste of time and money if the product idea is a set goal from the start! In that case, rather use the resources to validate whether that product is a good innovation bet or not. If you actually also want to explore an area more broadly, set up a separate project for that.

FORWARD, BACKWARD AND CIRCULAR

Another essential part of the design thinking methodology is that it is a circular process. In a circular way of thinking, it is perfectly fine and even recommended to move backward in the process when you learn more about the topic of your particular innovation project. However, the problem here is that the food and beverage industry has very linear processes due to many reasons and hence doesn't really encourage moving backward. It is all about moving forward and ensuring that all deadlines in the standard process, from the idea to the launch, are met. The most important reason for the linear mindset of the food and beverage industry is that most retailers use launch windows to handle their assortment. If your company wants to launch a new product in a supermarket, there are typically only a few periods each year when it is possible to get a new listing. This is a bit similar to professional football players who are typically traded during designated periods, known as transfer windows. Everything in the entire innovation process needs to be timed to certain launch windows for the target customer. The launch windows can vary from retailer to retailer, and even more from market to market.

In this time period, there are many processes to consider. One that distinctly sets the food and beverage industry apart from others is the critical matter of **food safety and shelf life.** The food and beverage industry is heavily regulated to protect consumers. Companies must comply with strict safety standards and regulations to avoid legal consequences and to maintain their operating licenses. They must ensure that the products are safe to consume during their entire shelf life. The implication is that if the aim is to have nine months of shelf life for the product, it will, in principle, take nine months to determine whether the product will still be safe to consume. There are ways to accelerate the shelf life tests, and that is widely used in the industry, but it comes with a higher risk.

Safe to consume is also only one aspect of the shelf life test. The second is how well the product performs on taste and texture during the shelf life. This aspect is more difficult to accelerate.

The Product Launch Push

What happens again and again is that the complexity and linear gravity of the industry dynamics motivate companies to proceed with projects before they are confident that the product solves the consumer's need and delivers the intended product experience. The following is a commonly used statement:

> *We know that the taste is not 100% where we want it to be, but we need to finalize the recipe now, in order to get everything ready for the launch window in 6 months.*

To stop and go back to the drawing board can be a hard decision to make. Nevertheless, it can be better for a company in the long run not to launch a product that consumers don't want to buy, and that risks being delisted after six months on the shelf – especially if rushing the process was the main driver behind the launch. It requires foresight and extensive planning to work with the circular design thinking methodology, and it is advisable to use the time needed.

On the other hand, you also need to be realistic in acknowledging the constraints that the food and beverage industry has.

Sometimes you need to go slow, to go fast, and as difficult as it can be to manage both needs in the innovation process, it is an essential discipline to master to launch successful consumer-centric innovations.

MOVEMENTS AND MINDSETS

The Double Diamond,[1] developed by the UK Design Council in 2005, is a visual model of the design thinking process that is very broadly used and iterated into many *builds* and new versions across literature and consulting companies. A basic illustration of this may be something as illustrated in Figure 6.1.

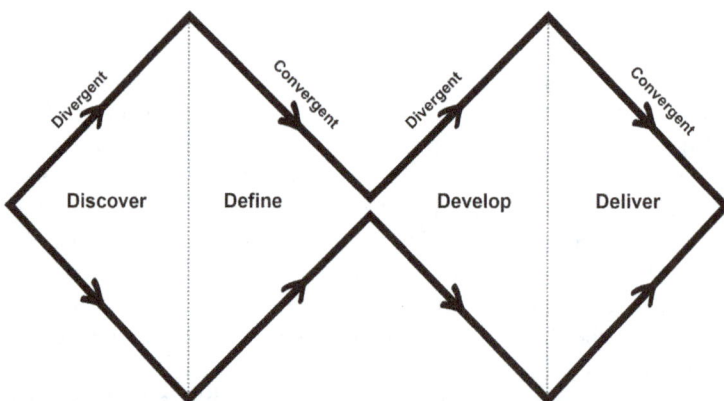

FIGURE 6.1 The Double Diamond. (Re-visualized by the author, based on the work of the UK Design Council.)

This balances divergent thinking, which expands possibilities, and also convergent thinking, which narrows down solutions. The Double Diamond process has four phases, and it begins with a divergent phase called **Discover**, where research and user insights help to explore a problem space. Next is a convergent phase, **Define**, where findings are analyzed to frame a clear problem statement. The **Develop** phase is once again divergent, encouraging ideation, prototyping, and experimentation with multiple solutions. Finally, the process returns to a convergent phase with **Deliver**, where the best solutions are tested, refined and implemented. The Double Diamond is an established and widely used model that provides a strong foundation for design thinking.

However, with a design thinking mindset, I have chosen to iterate and refine it by expanding the process into a more detailed, 8-phase model. This model will combine the iterative nature of design thinking with the structured progression of a gate process, creating an innovation framework, specifically tailored to the uniqueness and complexity of the food and beverage industry.

But before we go into the details of the 8-phase model, let's take a closer look at the meaning of divergent and convergent thinking. It is important to distinguish between the **movement** toward something broad or narrow and the **mindset** individuals naturally turn to during an innovation process. The two terms are not the same. The movement is an intentional process, while the mindset reflects how people naturally think and act within that process. The **divergent** movement is about expanding possibilities by gathering and exploring a wide range of information, ideas, and perspectives. It encourages openness and curiosity, ensuring that nothing is dismissed too early. A divergent mindset embraces this approach by welcoming all inputs, focusing on possibilities rather than limitations. Instead of questioning whether something is relevant or feasible, say:

Yes! Let's include this and build on it.

Some individuals naturally think this way, while others may find it challenging to stay open without filtering or judging information too soon. In contrast, the **convergent** movement focuses on narrowing down and making sense of the gathered information by organizing, categorizing, and synthesizing it. A convergent mindset helps to turn insights into structured assumptions, refining information and shaping it into something meaningful within the project's context. Here, the approach shifts from openness to critical evaluation, prompting questions like:

"Does this idea align with what we are trying to solve?", or "Is this feasible to produce?"

While some people thrive in this structured phase, others may resist committing to a direction, hesitant to let go of possibilities. An important factor in managing the innovation process is ensuring the right people are involved at the right stages. It is essential to be aware of how different mindsets influence the process, avoiding the mistake of assembling a team of entirely divergent thinkers during a convergent phase, or vice versa. While both mindsets should always be present, the balance should shift depending on the phase if possible.

During divergent phases, when exploration and idea generation are key factors, the team should lean toward open, exploratory thinkers who embrace ambiguity and possibility. However, in convergent phases, when decisions need to be made and ideas refined, the team should be weighted toward structured thinkers who can synthesize the information and drive the action. The goal isn't to exclude anyone, but to create a deliberate balance, ensuring the team's natural strengths align with the needs of each stage.

8 ESSENTIAL PHASES

With a solid understanding of when to adopt a divergent mindset and when to apply a convergent one, let's explore the 8 phases of design thinking that can serve as an innovation framework in the food and beverage industry:

1. UNDERSTAND (**Get It**) – Divergent movement
2. DEFINE (**Frame It**) – Convergent movement
3. IDEATE (**Solve It**) – Divergent movement
4. SELECT (**Choose It**) – Convergent movement
5. PROTOTYPE (**Build It**) – Both movements, but mostly divergent
6. TEST (**Show It**) – Both movements
7. EVALUATE (**Think about It**) – Convergent movement
8. EXECUTE (**Do It**) – Both movements, but mostly convergent

The flow of the eight phases that combine the divergent and convergent movements can be illustrated like waves, as shown in Figure 6.2.

The iterative nature of the process is indicated by the arrows going both ways in the model. In reality, there could be many more arrows applied during the entire process, all depending on the nature of the project and the time and resources available. The prototype and test phases are interconnected, and where most iterations are recommended.

THE SURFER AND WAVE METAPHOR

When discussing the phases as waves of expansion and contraction, along with the relentless need to reflect, learn, and adapt your solutions, it's tempting to fully

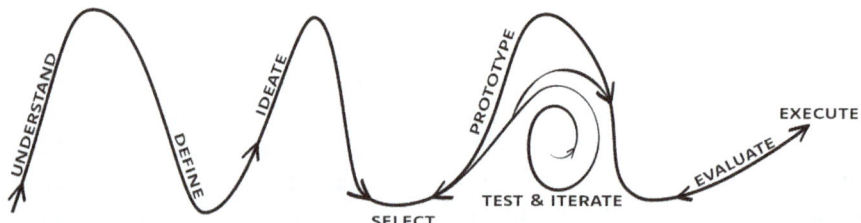

FIGURE 6.2   The 8 Innovation Phases.

embrace a surfer and wave metaphor. To illustrate what iteration is about and why it is important, imagine that you are a surfer and have had a great ride on a wave. Your body learned a bit more about balance and position when hitting the wave. Returning to the sea, your body is wiser and will use that learning when hitting the next wave. A skilled surfer watches and "understands" the ocean, analyzing the waves to prepare for the ride. Not every wave is a suitable wave, just as not every problem is a key one to solve. When a surfer catches a wave, balance, timing, and position are important and based on how the wave behaves. If the surfer falls during the ride, they swim back and catch a new wave. Surfers' experiences with one wave teach them what they need to adjust to better catch the next wave. It's a continuous learning process.

To succeed in innovation, you need to adopt a "Surfer Mindset."

As a quick teaser, here is a brief overview of each phase. More detailed chapters on all the phases will follow later.

UNDERSTAND (GET IT)

It's always beneficial in life to have a clear understanding of what you're doing and why you're doing it. It is the same in the food and beverage innovation. However, starting a project can be a bit of a chicken-and-egg situation. The *Understand phase* is all about the deep insights into human behavior and the motivation behind it. What are the emotional and functional needs consumers have in the context of food and beverage consumption, what barriers do they face, et cetera. Consumer insights like these are important! It is also about understanding the strategic position and capabilities your company has or potentially should have. You need to understand what to understand, why you should understand it, and how to understand it. Read that sentence again and reflect … – do you get it?

- Understand what to understand = What are your company's strategy and capabilities?
- Why is it important to understand = What would the consumers like to buy?
- How to understand = How to acquire consumer insights?

If your company already has a clear strategy, use that to guide what area of human behavior you need to understand better to develop new brilliant products they will love to buy. And as previously discussed, you might already have an idea of what you, as a company, would love to sell. If it turns out to be a case of mutual love, you could be catching a wave of opportunity.

It could also be that, in the process of obtaining consumer insights, you realize that it might be beneficial to adapt the company strategy or product constraints to better meet identified consumer needs. If you don't have a clear strategy as a company, the first project will be to determine that. In the ever-changing global market, consumers have diverse needs and preferences. They are not only seeking price-friendly fuel to sustain life. No, they also seek food and beverages that will provide unique taste experiences, convenience, options for health-conscious and environmentally-responsible products and much more. The essence of design thinking in a food and beverage creation context is to ask questions about motivations and barriers within F&B consumption. It is super important to really understand what drives the consumption of specific product categories.

Breakfast Example:

Try to ask why a person should have breakfast.

A possible answer could be: *"Because I need fuel for the day ahead."*
Why is that important?

The answer could be: *"Because I have important stuff to do at work and want to be able to perform."*

Ask teenagers the same type of questions, and the answers might be quite different.
Why do you have breakfast?

The answer could be: *"I don't, I normally skip breakfast."*
Why do you skip breakfast?

The answer could be: *"Because I like to sleep as long as possible and don't have time to eat breakfast."*

The point is that different people have different motives. Understanding these underlying motivations in the context of consumers' food and beverage choices is crucial in driving a successful innovation. Hence, consumer and market research are essential enablers in the *Understand phase*. However, before you begin an innovation project and enter this phase, a clear pre-scope will need to be formulated. More on how to approach that will follow later.

DEFINE (FRAME IT)

The define phase is where insights turn into focus – framing the project into a clear and actionable scope. After exploring consumer motivations, behavior, needs, trends, market, and competitor insights based on a project pre-scope, it is about organizing the findings into structured opportunity spaces that provide clear directions for innovation and set the stage for creative solutions. The define phase will result in a final project scope.

IDEATE (SOLVE IT)

The next phase may seem a bit more straightforward and a phase that most people really enjoy. They actually enjoy it so much that one of the main issues in the industry is that developers often jump to it before spending time on the first two phases. As stated at the beginning of the book, it is problematic to start with the solution without knowing what you are trying to solve. There is a large consensus that creating ideas is the easiest part of the innovation process, and to some extent, it is true. It is easy to create a lot of ideas! 100 ideas! Sure, no problem …, and with the help of AI, the number of ideas can easily get sky high.

But creating ideas that have the potential to become a successful innovation: that is difficult!

"80% of all launches fail!" This commonly cited "statistic" refers to the broader failure of new product launches in FMCG. Whether the figure is in fact 80% is hard to know, as it is not a fact-based statement. Nevertheless, the figure is high and there is a high risk that all the efforts of an innovation project will end up as a failed launch. So how can that risk be mitigated? First of all, the idea must solve a real problem for consumers. To make that possible, the *Understand phase* and the *final scope* provide the essential foundation for generating relevant ideas. Even if an idea addresses a genuine consumer need, that alone doesn't guarantee a successful innovation. If it doesn't meet a real need, it is almost certain the product will not achieve commercial success — meaning it may fail to generate sufficient sales, market traction, or profitability.

Most products that are unsuccessful will be removed from stores. They will be delisted because too few consumers buy them, or don't buy them frequently enough. Delisting can, however, also happen for other reasons, like retailer assortment prioritization. In this case, the retailer will need the shelf space for other products they prefer to try instead. Whether a product is delisted or not also depends on the size and influence of the company. The strength of the associated brand also matters. It is particularly challenging for smaller manufacturers to compete against larger, well-established brands. A broad view of unsuccessful innovation includes products that may still be on the shelves but aren't profitable.

SELECT (CHOOSE IT)

This phase is a hard nut to crack! You can't simply prototype and test all the ideas from the ideation phase and then choose the best performing prototype and work with it. You have to make hard choices, and it is about assessing the fit with the defined opportunity space and other important lenses that can be used as filters. Taking the time to have in-depth conversations about this is essential. As the design thinking mindset is iterative, the select phase can be revisited at a later stage if you realize that it could make sense to look at the ideas again.

PROTOTYPE (BUILD IT)

The focus now shifts to building the selected ideas into prototypes. The purpose of a prototype is to make the idea tangible enough to share with potential consumers and to gather their feedback. It is about showing what the product is about, and for what occasion it is intended. It can be just like packaging mockups and text, but at some point, it is also about food and beverages to be consumed and evaluated! How do you create a prototype with a balanced taste profile that can be scaled up to run on a production line without significantly altering the taste and texture? How do you serve it? Do you use anonymous packaging, or packaging that also sets expectations for the product experience? Prototypes can come in many levels of fidelity, and there are a lot of important factors to consider when creating them.

TEST (SHOW IT)

In this phase, the prototypes will be shown to understand if the ideas resonate with the target audience. This can be tested in many ways, and it is a core discipline in the design thinking methodology. The test phase represents the second stage of integrating consumer research into the innovation process. You can ask consumers questions via online questionnaires, conduct in-person conversations with them, have them try the products, or even use neuroscience to measure how the taste experience is connected to emotions and memories. Based on all the feedback received from the test phase, you once more need to assess the results and learnings. Perhaps you have to go back to one of the earlier phases and iterate some of the ideas. It will most likely be an iteration in the prototype phase, but also in the select phase, or the ideation phase can be revisited.

EVALUATE (THINK ABOUT IT)

The final goal is, of course, to have clarity on what specific idea has the best potential to become a successful innovation for your company. It is once more a difficult decision to make and be sure to do thorough analysis of all the factors to consider, before making the call. Think and talk about it before taking an informed decision. In some cases it can make sense to go back to the understand phase, do further research and potentially restart the process, based on new learnings.

EXECUTE (DO IT)

The decision has been made and now the process is moving from a place where it is okay and even advisable to move backward in the process, to a place where it is not. The circular mindset is to be parked, and the linear way of working is the goal. You need to do it! Set all the processes in motion to finalize the product recipe, sourcing the ingredients and packaging needed. Scale up and validate that the output from the production line matches the reference prototype and target brief. Check the shelf life. Design the packaging with spot-on communication. A final check with the target audience in a validation test – last chance to pull the emergency brake! Get listings at the retailers, plan marketing activities, and **LAUNCH!**

THE FIVE LENSES OF INNOVATION

Sometimes it can feel like there is no new innovations to be invented! Or, at least, it will take too much effort to go after something new to the world. It will be too risky, or you may assume the competitors are way ahead already. The discussions within companies on these topics are "on repeat."

Past Failures and Fear of Repetition:

Statement:

Let's try with this concept again, we have all the recipes and can just press play.

Push back:

We have already tried it, and it was a huge failure! If it didn't work then, why would it work now?

Risk and Uncertainty:

Statement:

I really believe in this idea, I think it will be a breakthrough product for us.

Push back:

We can't afford to take risks with something this unproven.

Market Saturation and Competition:

Statement:

Should we make a 'me too' of the competition and get a share of the category?

Push back:

Why bother entering a fight we can't win?

And the list goes on … It is typical conflicts between radical and incremental ideas, perceived business potential, and alignment with core company values. These are not easy discussions! They can be hard, and as always, different stakeholders have different leverage, views and agendas. During the phases of the innovation process that require convergence and agreement within the team and among stakeholders, it is valuable to examine learnings and ideas from different perspectives. The goal is to provide a structured framework that facilitates discussion and supports decisions based primarily on data and well-founded arguments – minimizing the influence of personal opinions from stakeholders who may lack the core team's depth of research knowledge and consumer empathy.

I suggest using five different perspectives: **The five lenses of innovation.**

The first three are the ones typically mentioned in design thinking literature.[2] If you can tick boxes in all three lenses, there is a good indication that you have identified a sweet spot for innovation.

The lenses are the following:

1. Desirability
2. Feasibility
3. Viability

In addition to these three lenses, it makes sense to also incorporate the Corporate Social Responsibility (CSR) policy of the company and add two extra lenses, namely:

4. Sustainability
5. Responsibility

One can see it as a checklist of the key critical questions to ask when investigating which ideas seem to be the most attractive to take further in the process. **How you, the business you operate or work for, prioritize the weight of the five lenses can, of course, be very individual.**

The five innovation lenses and some examples of questions to ask:

DESIRABILITY

The lens of "will the consumers buy this product, and will it meet or even exceed their expectations?"

Typical questions to ask:

- Who could the target audience of your project be?
- Is the project seeded in a food or behavioral trend? Describe and validate the trend.
- Why would the project resonate with the emotional and functional needs of the target audience?
- What is the occasion? When, where, and why would the consumer use the product during that occasion?

FEASIBILITY

The lens of "will we be able to produce this product?"

Typical questions to ask:

- Do we have the production capabilities to produce this type of product?
- Do we have the necessary product development skills to create a recipe that will give the desired product experience?
- Do we have the packaging that will provide the desired functionality and product expectations, and experience?
- Will we be able to source the needed ingredients, and would they all be allowed in the factory? (E.g., due to allergens, et cetera.)
- Will we be able to produce the product with a shelf life that meets or exceeds market standards?
- **N.B.** If the answer is negative to some of these questions, would it be possible to get it produced by another company?

VIABILITY

The lens of "will it be a good business?"

Typical questions to ask:

- Do you have a brand that fits with the proposition and target audience?
- What is the potential production cost of this type of product, and how does it compare to the competition? Essentially, can it be produced at a cost that allows for an acceptable profit margin for the company, while still offering a selling price attractive to consumers?

- Will the product have a competitive advantage, based on its unique selling points? (USPs)
- Will the product idea have a major risk to cannibalize an existing product in your company's current range?
- What will the estimated size of the opportunity be?

SUSTAINABILITY

The lens of "will the planet benefit from this product?"

Typical questions to ask:

- Will we be able to produce the product with low or reduced CO_2 emissions?
- Could we use ingredients from food waste streams?
- Will we have packaging options available with low or reduced CO_2 emissions?
- Will the packaging be recyclable?

RESPONSIBILITY

The lens of "will it bring value to society?"

Typical questions to ask:

- Will the products be safe and healthy to consume?
- Will the production benefit the local communities? (E.g., an increased workforce; the use of local ingredients; or to bring health and nutrition to populations with low food availability.)
- Will the desired communication of the proposition be credible? (E.g., maybe the product claims to be healthy. Make sure that is the truth.)

The ideas that can survive the relevant critical questions have the potential to become successful innovations. That is true, however, it is also important not to let critical questions disrupt the flow of the process and risk the possibility that interesting ideas are not tested, due to the uncertainty of the quality of the answers that can be provided at an early stage. It is a balance when to ask these questions. The questions about desirability are, of course, essential in the Define phase, but don't reject an idea at that stage if all the questions around feasibility and viability are still unclear.

Further down the line in the Evaluation phase, it is very important to have good answers to all the relevant critical questions from all five lenses, to make a qualified decision to take the product idea to the Execute phase. Many product launches in the food and beverage industry are based on gut feeling, experience, own preferences, et cetera. Product launches with no consumer data or other research to support the bet. It is very common, and not necessarily a wrong approach! If you are a smaller company with limited resources, it might be the only way. If you are passionate about

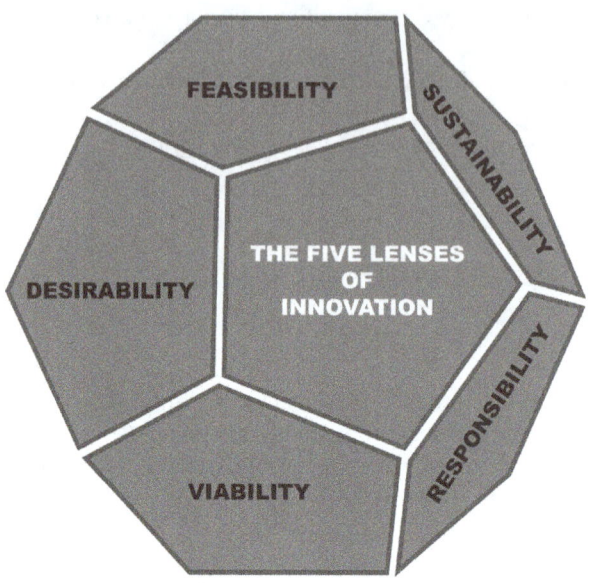

FIGURE 6.3   The Five Lenses of Innovation.

your idea and really believe in it, it will in some cases also be a success, providing the following conditions:

- You are launching at the right time and place.
- You have already developed a delicious product that suits the target audience.
- You hit the consumer's functional needs or emotional needs, or both.
- You can get it listed and make it available for the target audience.
- You can inform the target audience that a new product, specific for them, is out there.
- The expectations you communicate will be fulfilled with the product experience.
- The price is right.
- And we could go on …

You will indeed be very lucky for all these factors to be aligned without asking consumers what problems they actually face that are not already solved by competitors, or what they are willing to pay to solve them.

Consumer research is an essential part of the design thinking methodology. In the Understand phase, it is the core foundation, and later in the Test phase, it is an indispensable validation of whether you actually "got it right." The iterative nature of **design thinking** will prompt the consumer to understand the product repeatedly throughout the innovation process. Therefore, if you want to embrace design thinking in your innovation process, consumer research in some shape or form is where to start. It doesn't have to be a very fancy and expensive research to get started. Generally speaking, I would say, it is better to do something rather than ignore it.

Ask the consumers:

I have worked in companies with very different levels of resources. In some of the companies, we had very few resources to pay for research companies to investigate topics and compile reports or recommendations. We would look at what was available of free public research, and build on that. We would look at food trends and competitor launches. And build on that. What about testing prototypes with consumers? That didn't happen! We aligned and tasted the prototypes internally and launched them whenever there was consensus around the potential of the product. It always helped if the CEO and the sales force understood and liked the product. If the sales force, even recently, had a customer request similar to the product idea, things could move fast.

Would it have been hard to ask consumers to take a look and give feedback? No, it would have been pretty easy, but such a culture and mindset were not present at the time. It is better to get some sort of unbiased sense checking of your idea with some consumers, than none whatsoever.

In other companies, I have had the privilege of having consumer research available, and being able to perform proper targeted consumer research internally, or to hire specialized companies to conduct it for us. Testing prototypes with consumers? Yes, certainly. Further validation tests? Yes, certainly. Possibilities that I would have given my right arm for, when working in the smaller companies. In big companies, possibilities that accumulate huge amounts of data are sometimes, sadly, not used, or not used properly.

How companies go about acquiring consumer research can be many.

As I see it there are two basic rules:

1. If you don't have a lot of resources, be creative, be agile!
2. If you have a lot of resources, be mindful to utilize them with care and thoughtfulness.

It is fair to point out that consumer research is unfortunately not a guarantee for success! Many failed launches have gone through thorough research, but, for some reason, they have not succeeded. The explanations for a failed launch may be many. After a failed launch, take a look through the five lenses and see if it can provide you with insights to learn from – before you start the next innovation project.

DESIRABILITY

- **Consumer preferences**
  Misunderstanding consumer preferences and market trends can lead to products that don't meet customer needs or expectations. Or maybe by the time you launch, the consumers have moved on.
- **Product taste**
  If the product doesn't meet the consumer's taste expectations, it will likely fail.
- **Ineffective marketing**
  A lack of, or poor marketing strategies, can prevent a product from gaining traction. This includes insufficient advertising, the wrong target audience, or unclear messaging.

- **Bad timing**
  Launching a product at the wrong time can result in poor sales, such as introducing a seasonal product too late or too early.

FEASIBILITY
- **Product quality**
  If the product has quality issues, it will likely fail.
- **Packaging issues**
  Packaging that is inconvenient or unattractive in some way can lead to product rejection.
- **Regulatory and safety issues**
  Products that fail to meet safety standards or face regulatory issues like claims that are not allowed, can be pulled from the market.
- **Logistical problems**
  Issues in the supply chain, distribution, or production can cause delays and reduce product availability, leading to failure.

VIABILTY
- **Price point issues**
  Setting the price too high or too low can be a problem. High prices may be seen as unjustified and too expensive by consumers, while low prices might mean too low margins for your business.
- **Competition**
  Entering a saturated market or being outperformed by competitors can lead to failure.

SUSTAINABILITY
- **Recycle issues**
  Packaging that is not environmentally friendly can lead to product rejection.
- **Greenwashing**
  Claims on the package that suggest sustainability, but not in a trustworthy way, can cause consumers to ask questions and lead to negative publicity.

RESPONSIBILITY
- **Public health**
  The product targeted children's daily consumption, but with a high sugar content, causing families to deselect it.
- **Child labor**
  If it turns out that some of the ingredients were sourced from an origin associated with child labor, it is very bad – not just for the launch, but the entire reputation of the company.

If you work, or have worked for a company with failed launches in the baggage, remember that all companies have some or other struggles.

Understanding these reasons can help to avoid similar pitfalls and increase the chances of a successful product launch in the competitive food and beverage industry.

COMPANY STRATEGY AND BRAND POSITION

Before you embark on the product design journey and start a product innovation project, you need to decide what the scope of the project is. You should give some direction on what to investigate and possibly create ideas against. These directions should be based on a deep understanding of the company strategy and the specific brand position.

A company strategy must be in place!

If not, start with a thorough market analysis of the external environment, including market trends, the competitive landscape, consumer or customer needs, future foresight, and so on. It is not the aim of this book to focus on how to develop a company strategy, but you should actually use the same design-thinking mindset and process, as if it were a product innovation project. Put together a diverse team and capture the essence of what your company should be about, and why it will make a difference for the consumers! Strategy is not a constant. So even if your company is well-established and has a strategy in place, it also needs to be regularly adjusted. The most companies do that, but some might also think that their strategy doesn't need adjustment. There is a famous quote in business management:

Culture eats strategy for breakfast[3]

It refers to the internal company culture and how it is essential that the strategy of a company is embedded in the mindset of the employees. This is necessary so that everyone will understand the strategy and work as a team to accomplish the goals connected to the strategy, but culture often has a stronger influence on the success of organization than strategy alone.

The ideas discussed above are very true, but how to run a successful company is a whole new topic! Inspired by Drucker's quote, but viewed through the lens of the design thinking methodology, I will reframe it within the context of consumer centricity:

Consumer behavior eats strategy for breakfast!

To underscore that no matter how robust or well-crafted a business strategy is, it will struggle or fail if it doesn't align with actual consumer behavior, at the point in time where it is executed. Hence, future foresight is highly relevant to incorporate into strategic work, and this topic will be explored in greater detail in a later chapter.

Working with the strategy, you normally start with formulating the vision, the mission, and if the core values of the company are not clear, it is essential to formulate those too.

Vision, Mission and Values

The *vision* is the company's aspirations and desired future state, while the *mission* statement defines the purpose of the organization and its core reason for existence. *Values* are the foundational principles or beliefs that guide a company's culture, behavior, and decision-making. They reflect what the company stands for, such as integrity, innovation, sustainability, and responsibility. The values will also help to frame and formulate the CSR policy.

As an example, if you have a company like *Reduced*, sustainability and responsibility are important parts of your values and mission, and will have a high impact on what type of innovation projects it would make sense to initiate. It could also be in the values where it is defined what type of innovations are the main objective of the company. Most companies would probably say that a good mix of radical and incremental innovation is what they aspire to accomplish. Who wouldn't like to innovate something groundbreaking new to the world? The most companies and their CEOs would be proud and happy if they succeeded with that. On the other hand, most companies and their CEOs also have ambitious goals and financial targets. Short-term goals might overshadow the more long-term goals.

Goals and Objectives

It is important to have clear, measurable goals and objectives to support the company's vision and mission. These may include financial targets, market share objectives, or growth milestones. Equally important is the allocation of resources to fit the strategic priorities. How much of the company's resources will be allocated to innovation?

Is the focus on short-term wins, or rather the long-term R&D and potential radical innovations?

Brands

There is a big difference between working with an existing brand and creating a new brand. It goes without saying. In existing companies, the brand position is usually already well established. Brand positioning is how a brand is perceived in the minds of consumers in terms of its unique value proposition, product attributes, benefits, and personality.

What does the brand stand for and how does it differ from the competitors is important to know. You need a distinct visual brand identity to differentiate it from the competitors and to resonate with the target audience.

Key brand positioning elements should include the target audience as well as the brand identity, and the brand promise.

- **Target Audience:**
  One should identify and understand the specific segment of the market that the brand aims to target and serve.

- **Brand Identity**:
  This includes the distinct identity for the brand that resonates with the consumers to evoke emotional connections.
- **Brand Promise**:
  The position of the brand and what it promises are important, e.g., the offering of premium products, convenience products, low prices, a sustainable choice, et cetera. This is the unique value proposition that sets the brand apart from competitors and provides a compelling reason for consumers to choose it over alternatives.

With a clear company strategy in place and a well-defined brand, one can start to ask questions and make assumptions to be tested in the Understand phase. A brand strategy must also be in place or parallel with the product innovation! We will get back to marketing and the brand strategy in the Execute phase.

Startup Company and Brand Building

If you're in the startup position and unsure where to begin, it's generally wise to prioritize your company strategy. If you haven't developed a brand yet, you can build it alongside your first product. Just be careful not to create a brand that's too narrow or limiting. Ideally, it should leave room for a future range of products.

Global or Local

Depending on the company's setup and its vision and mission, it will be natural to include your ambition in terms of the markets you plan to have as core markets. Some companies are very strong in the market that they originate from, and some are stronger in exporting the products. If you can achieve a global hit, it is groundbreaking for a company. My experience at a global FMCG company that owns renowned brands has shown me how significant this is within the organization, when it comes to turnover as well as corporate pride.

One doesn't become a global player without challenges! There are many things to consider in terms of how to get distribution and to manage the logistics, have sufficient shelf life, and so on. Most importantly, however, is to ensure that the product meets consumer needs and taste preferences across markets. When the taste experience is solved for a global audience, it is important to communicate it in a way that is clear for all nationalities. Enabling design thinking and tailored consumer research and testing can help to accomplish this.

My Global Attempt Story

After finalizing the food architect program in 2011, I struggled at the B2B company to be able to implement some of the great design thinking ways of working I had been exposed to. I remember trying to advocate for customer interviews within the NPD department to detect better ways to serve them. Despite having the NPD manager's support, I met fierce resistance, and the design thinking never took off. Pretty soon, I decided to educate myself further in the design thinking methodology, and luckily, I was the owner of a small spice company: Aroma Spices.

At the time, the setup was very classic in terms of assortment and innovation. It was a local market setup, and the products were sold online. The webshop and the setup for the business model were a bit unconventional, as it was through one of the large newspapers[4] in Denmark. The newspaper had a special section for its subscribers, where they could buy exclusive stuff. The food section in the newspaper was a strategic bet, and the subscriber webshop had a range of food specialities.

My company was the premium spice brand, and because it was a part-time company, I roasted and blended the spice blends in a miniature kitchen setup at my in-laws' large double villa, during evenings and weekends. My father in-law was kind enough to pack and ship the orders. The setup had been working fine for a couple of years, but to be able to pursue the wish to educate myself in what I learned during the program, I canceled my agreement with the newspaper and started my first attempt to hit the global market with a "banger!"

In November 2011, I started the project: The world's best curry powder project! And yes – I started with a product idea! My inspiration was Tabasco Sauce, a great product that I encountered all over the world. One product, for a global market!

And yes, going back in time with the knowledge that I have acquired since, I might have done several things differently. To be fair, I made many mistakes, and it was, at times, a struggle!

And yes, it was a very bold title and also widely misunderstood, as the "world's best" didn't refer to the product, but the project! Well, it still stands, I guess, as I haven't heard about any other curry powder projects, so far … Basically, I just wanted to create the "Curry Powder" which people would recommend to each other, if they were asked to point out a really good one! If you lived in Copenhagen, London, Tokyo, or anywhere else, for that matter – not a range of curry powders. One curry powder to be sold globally!

My Design Process

I started by testing a lot of the existing competitor curry powders on the market – and profiled the taste and cooking functionality. With these taste profiles in mind, I traveled to Kerala in India – the origin of some of the key ingredients in a curry powder. Here, I discovered the authentic tastes of the southern Indian cuisine and sourced several special spices for the project. When I returned to Denmark, I started the real development work, compiling a lot of different recipes.

I documented all the steps of the project on Facebook and got people involved. An international tasting panel of chefs and foodies was assembled, and samples were sent all around the world. The feedback was an essential part of the development process. In August 2014, I launched the product, CP44, and in 2015, I won a Great Taste Award in the UK, and before I sold the company in 2016, it was distributed in the UK, US, DE, AT, CH, EE, and even a first batch was shipped to the homeland of yellow curry powder, India. Sadly, the global reach failed in the long run, but it is still on the market in DK to this day (2025).

FIGURE 6.4 CP44 Curry Powder. (Pack design by the author of this book. Courtesy of [www.karlsenskrydderier.dk], used with permission. Picture taken by www.christianpetersen.dk.)

TEAMWORK AND COMPANY (SIZE AND TYPE)

For a bigger design thinking process, you need to put together a great team. The number of people and their qualities all depend on the company that you work for.

It is crucial to put together as strong a team as possible and to lean into the innovation process. In the book, *Winning at New Products*[5] by Robert G. Cooper, he shows that the best innovators are those companies that allocate sufficient resources for the innovation process and put together well-balanced and diverse teams to deeply focus on the innovation project. Including clear leads to handle the project from start to end.

Small companies will have fewer resources to choose from, while bigger companies will have more. Small companies and startups will work with the resources available. That goes without saying. A single skilled individual can make a big impact, taking on multiple roles and approaching problems from different angles. I also recommend reaching out to the network in these cases, to get more perspectives on your innovation work. Most innovators are happy to help, and if need be, have them sign a non-disclosure agreement (NDA). As companies increase in size and market share, resources to optimize the innovation process and allocate resources will follow. Or at least, should follow.

It is not uncommon to see that growing companies will put more resources into the marketing and sales force but keep the same level of innovation. If you want to survive in the long run and grow as a company, I recommend prioritizing and boosting the resources for your innovation efforts. The competition is fierce, and companies that hold back on innovation will be surpassed by those that keep pushing forward and launching exciting and relevant new products for consumers.

Throughout the eight essential phases of the innovation process, I will outline key initiatives that help build a strong foundation for making informed decisions on what to develop and launch. A successful innovation process requires a variety of roles in a team, each contributing to different aspects of the journey. Keep in mind the importance of balancing divergent and convergent mindsets, ensuring that the right people are involved at the right stages, with a team weighted toward the dominant mindset that best supports the movement at that phase. Some of the essential roles in the design thinking innovation process include:

Project Leader

When you start up a new project, it is key to success that you appoint an overall Project Leader. Be sure that the leader has a mandate to make decisions and a budget to navigate within. The project leader has the overall responsibility to organize the project into the eight essential phases and ensure it is running according to plan. In larger companies, it will be natural to appoint an innovation manager. In smaller companies, the choices might be fewer, with perhaps only one or a few individuals for whom it would make sense to take the lead.

Core Team

It is advisable to create a core project team of 2–6 people. More team members can make it harder to make decisions. Ensure some diversity in the core team. It is important to have different views to make good decisions. A good mix could be:

- Project manager (Overview and direction)
- Product designer, or consumer scientist (Consumer lens)
- Product developer (Product lens)
- Brand manager, or innovation manager (Commercial lens)
- Process Specialist (Production lens)
- Sales Manager (Sales lens)

An overall list of roles and the competencies needed to leverage successful innovation can look really extensive. Each company has roles that blend with adjacent roles. The titles are also being used inconsistently, and new titles emerge from time to time. More details on the different roles in the F&B industry are to be found in the vocabulary section at the end of the book. Leverage diversity as a guiding principle when hiring people for these roles and responsibilities. Don't hesitate to involve external expertise when appropriate and possible during a project.

Task Leader

During the different phases of the project process, there may be situations when it may be fruitful to appoint a task leader. Again, it depends on the company size and the resources in deciding whether these task leaders can be internal or external people. The task leader is different from the project leader, as they only take the lead in specific phases of the innovation process, as the experts in a certain area related to the phase. It is, of course, not always possible to have dedicated individuals for each phase in all types of projects and companies, but this is an overview of the various positions that could be relevant. In the phases where key decisions are the essence of a phase, the project leader will also be the task leader, and it is important to include the core team in taking the hard decisions.

1. UNDERSTAND (Consumer scientist, Product designer, or an Innovation manager)
2. DEFINE (Project leader/Core Team)
3. IDEATE (Facilitator)
4. SELECT (Project leader/Core Team)
5. PROTOTYPE (Product designer or product developer)
6. TEST (Sensory & Consumer scientists, or Product designer)
7. EVALUATE (Project leader/Core Team)
8. EXECUTE (Brand/Innovation Manager)

JOKERS

To add additional diversity to a team, consider including some knowledgeable and creative *Jokers* – an outsider perspective! These individuals could be everything from experts from other industries, to anthropologists, or even artists, who provide unconventional thinking and creativity. They can challenge assumptions, ask naive questions, and share wild and creative inputs and reflections. Though not involved in the daily operations, they can spark breakthroughs by interacting with nearly any team member at strategic moments. Roberto Verganti[6] calls them "Interpreters" in the book, *Design-Driven Innovation*, and even though that book is about radical innovation, I still believe it is a valuable point in any project to get novel outside perspectives. Later in this current book, there is also a section about how to work with future foresighting. Here it is even more essential to include Jokers.

DECISION GATE MODEL

The larger the food and beverage company you work for, the more likely it is that it has a stage gate model. It can have many shapes and forms. The "original" stage gate model[7] is actually a full innovation process framework. However, in many companies, the main purpose of having a gate model in place is to ensure that there is a systematized way of assessing the projects and grant resources to the ones with the best potential to provide success for the company. The model can be shaped in many ways. It all depends on the type and size of the company and the innovation process journey it has been on.

When a gate model is worst, it doesn't bring ideas to life, it kills ideas!
Or even worse, it approves ideas with no real point of difference, based on an artificially high assessment of the return on investment (ROI).

In this book, the innovation framework is based on the design thinking methodology. A linear gate mindset versus the circular and iterative nature of design thinking can seem very counterintuitive, but they can and should work together. At its best, it supports the design thinking innovation, radical as well as incremental.

If a solid company strategy is in place and the goal is to also work with short-, mid-, and long-term innovation projects, it should be fairly simple to set budgets aside for all three different time horizons. Mid- and long-term projects don't have to end up in a product new to the world, as previously stated. In the food and beverage industry, it rarely happens. What is essential, though, is that the process supports the possibility of radical innovation and that the team is allowed to work accordingly,with the freedom and resources needed to explore bold ideas. Avoid a decision board of only high-positioned members of the company. Include diversity in the decisions. Include employees with a deep understanding of the project. The problem is that in the higher positions in a company one is more likely to become a bit detached from reality, in the sense that you will be presented with oversimplified versions of the potential and implications of a project.

Sometimes the best decisions come from relying on the project leaders and core teams to deeply understand the problem, while the leaders focus on strategic alignment.

Decision Gates

As illustrated in Figure 6.5, I suggest having three decision gates to support the innovation process.

Gate 1: Before deciding what type of innovation project to initiate, approve, or decline.

Gate 2: Before creating new ideas, approve or decline the final project scope.

Gate 3: Before deciding to execute, approve or decline the specific idea.

PRE-SCOPE AND DECISION GATE 1

Before diving deeply into exploring and understanding an area of interest, you first need to identify and formulate a pre-scope of what the area of interest is! For a food and beverage company, you obviously have some ideas of what product category would be interesting to explore for future growth opportunities. The ideas should be based on your company strategy, or a movement in the market that seems to grow and have potential. As mentioned earlier, some very tangible ideas might already be talked about, that might, or might not, be the next new product that the potential consumers will be looking for – consciously or unconsciously!

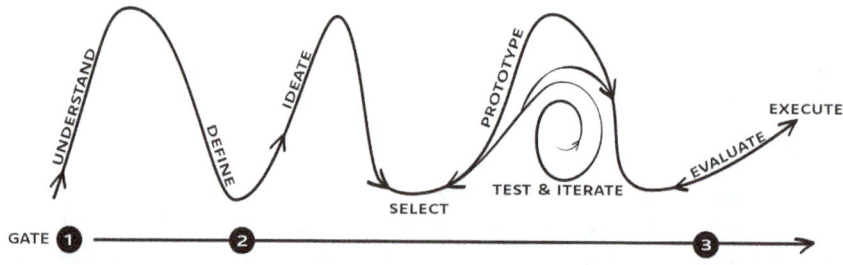

FIGURE 6.5   The 8 Innovation Phases and 3 Gates.

If stakeholders have already decided to move forward with a specific product idea based on "inside out" reasons, like: We have the production capacity free, it's a cool product, and by the way, our CEO loves it – and yes, it happens … – then it's worth stepping back and asking whether it really makes sense to initiate a full innovation project. If the pre-scope is simply about making the idea fit rather than questioning its actual desirability, the approach should be reconsidered. It is important to be mindful not to approach problem identification in the same way when a specific product idea is already the goal from the start. Don't pretend you are exploring broader consumer behavior without a solution in mind if that's not the case. Instead, use your resources more efficiently to understand whether this particular idea resonates with the consumers and how it might need to be iterated to truly solve a relevant problem.

Let's go back to the 8-phase process.

If you already have a product or concept idea in the pipeline, you are basically in the Prototype and Test phase. Depending on how the product idea was generated, you might have knowledge of what assumptions it was based on.

What target audience it is aiming for, and what problem it will solve for them. What are the drivers of preference this audience has regarding the product experience, and so on … If it is the case that you have data and clear assumptions from an Understand phase, you can create some prototypes and start to test the assumptions with the target audience. In many cases, this will be a line extension, and you can extract a lot of data from the success or challenges from the existing category products on the market, whether it be your own and the competitors'. If it doesn't work after tests and iterations or something feels forced or biased, raise the red flag, as it might be that you need to stop the project and perhaps start again from a more fundamental area of interest.

Utilize a more correct way to work with product design thinking. The standard approach from a design thinking point of view would be to avoid too firm a set of ideas at the early stage of the process. Try to keep the questions open and not too specific regarding product ideas. It could be useful to use questions like the following:

- What are the behavior and challenges connected to a healthy lifestyle within a specific target audience?
- Could our existing brand stretch into a new category or occasion? Maybe one can even expand the target audience. Be specific regarding the assumption about what category, occasion, and target audience you have in mind.
- What potential could the higher focus on sustainable food choices in society mean for us?
- Could inflation have an impact on consumer shopping behavior, and could it create a need for new types of products to make dinner cheaper?
- Is there an untapped potential for the breakfast occasion for new types of products?

That said, I recommend not being too rigid about the specific formulations. Words and formulation are important, but from time to time you end up in a fight over very insignificant differences. With all that in mind, before you start a project, you need to pre-scope the area and write a project brief to provide the team that will lead the work with the necessary information and framework to initiate a targeted execution.

A Pre-Scope template could look something like this:

Area of interest: A concise phrasing of the area and identifying potential launch markets.

Potential category: Be specific or open-ended, allowing flexibility to pivot towards a radical innovation project, if consumer insights support that direction.

Essential questions: Questions about behavioral trends, product trends, occasion, et cetera.

Strategic importance: How does this project align with the company's vision and mission?

Brand and timeline: What brand is in scope, or could it be a new brand? What part of the pipeline is the project supporting? Short-, mid-, or long-term.

THE GOLDEN JUICE COMPANY

A fictive company to be used as an example of how the pre-scope could look.

Some background info:

- Company name: The Golden Juice Company (TGJC)
- Established: 1995, Herefordshire, UK
- Herefordshire Headquarters include Production, R&D, Marketing, Sales and Administration.
- Additional production site in Poland and sales offices in core markets.

THE STORY

Nestled in the hills of Herefordshire, renowned for its beautiful apple orchards, The Golden Juice Company began as a dream of a young couple who left the bustle of London, trading their corporate careers for the charm of the countryside and entrepreneurship. It was the summer of 1995 when Sarah and Tom Golding bought the old apple farm with 10 acres of land and over 300 apple trees. The couple began with the small-batch production already located at the farm, but it expanded into a fairly large company during the first 10 years. Today, The Golden Juice Company operates a "state-of-the-art" facility in the heart of Herefordshire, producing millions of liters annually, while staying true to its roots.

Their flagship juice product, **Golden Dawn**, remains a customer favorite. It is a nifty blend of a variety of apples, providing a flavorful, sweet, and tangy taste. The taste profile is accomplished by using the well-known cooking apple, *Bramley*, as one of the differentiating apples in the blend.

VISION

To be the world's leading provider of premium, natural beverages, inspiring healthier lives and creating a sustainable future for generations to come. The slogan is: "Golden Quality – Global Reach."

Herefordshire UK
Est: 1995

FIGURE 6.6   Fictive Company Logo.

MISSION

We are on a mission to deliver exceptional, fruit-based beverages to every corner of the globe. By combining our heritage with cutting-edge innovation, we aim to meet the evolving taste preferences of our consumers while championing sustainability, quality, and community engagement.

As an example, this is how a pre-scope formulation could look for the Golden Juice Company:

PRE-SCOPE: INNOVATION PROJECT – "NEW DAWN"

Area of interest: We want to understand how we can leverage the breakfast occasion with a new type of product in our three core markets: UK, DE, and NL.

Potential category: Smoothies in some form. We are open to exploring new types of ingredients for the factory and new production processes if needed.

Essential questions:

- What are the behaviors and challenges connected to a busy lifestyle, but still aspiring to be healthy?
- What potential target audiences would be in scope?
- How do their mornings look? We are particularly interested in the "on the go" moments.
- What are the usage and attitudes towards smoothies today?
- We have seen more and more veggies appearing in new products from the competition. Should we also go that way?

sthfldj ldfkjfrrf

TABLE 6.1

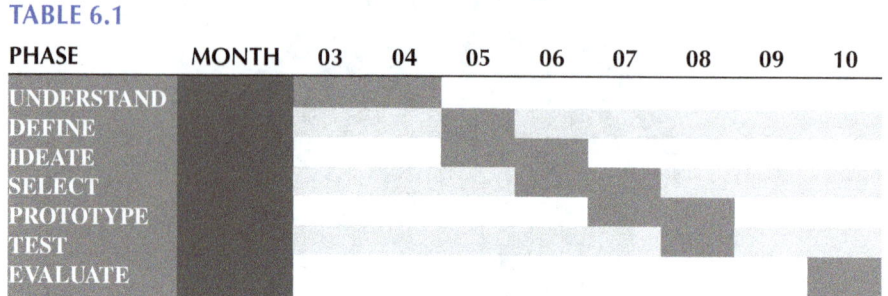

PHASE	MONTH	03	04	05	06	07	08	09	10
UNDERSTAND									
DEFINE									
IDEATE									
SELECT									
PROTOTYPE									
TEST									
EVALUATE									

Strategic importance: We want to help consumers live healthier lives and see a potential to create a new type of breakfast product for a target audience, who are too busy to prioritize having a "normal" breakfast.

Brand and timeline: The project is to support the mid-term pipeline of our brand, Golden Dawn, or a new sub-brand.

We suggest kicking off the project in January, with a project timeline looking like something like illustrated in Table 6.1.

As it is a mid-term pipeline project to potentially be launched in 2 - 3 years, the last two phases, execute and launch, will be determined in the Evaluation phase and taken to gate 3.

Suggested team and project leader: Here, they will provide a suggestion on who to head up the project, and what a team setup could look like. This is also including suggested external partners to hire and to collaborate with, along with a budget estimation.

DECISION GATE 1

In the decision gate 1, the purpose is to present the pre-scope and explain why this is an important project for the company. The goal is to gain alignment on the value and relevance of the project, ideally resulting in approval to initiate the next phase. The decision-makers may also choose to provide a mandate for specific areas within the project.

In the case of the New Dawn innovation project, the closing message was:

- We approve the innovation project and budget.
- We are open to exploring new types of ingredients for the factory, as well as innovative production methods.
- If new production equipment is required, the board is open to considering potential CapEx[8] investments.
- We support the use of our Golden Dawn Brand but are also open to alternative brand directions, if guided by insights.

Ready?

We are now finally at the entrance of the eight essential innovation project phases.

The first phase of understanding can feel like an overwhelming list of routes of what and how to investigate an area of interest, but it is important to clarify that it is just that. These are potential routes to enable a deeper understanding as a foundation to innovate against. What will be possible and meaningful to initiate will be very company- and project-dependent. Welcome to the first phase!

The understand phase.

NOTES

1　Design Council. 2005. The Double Diamond. Retrieved from www.designcouncil.org.uk/our-resources/the-double-diamond/
2　*Change by Design, Revised and Updated: How Design Thinking Creates New Alternatives for Business and Society*, Tim Brown & Barry Katz, Harper Business, New York, 2019. Page19.
3　The phrase *"Culture eats strategy for breakfast"* is widely attributed to **Peter Drucker**, a renowned management consultant and author.
4　https://politiken.dk/
5　*Winning at New Products: Creating Value Through Innovation* (5th ed.), Robert G. Cooper, Basic Books, New York, 2023. Chapter 3.
6　*Design-Driven Innovation: Changing the Rules of Competition by Radically Innovating What Things Mean*, Roberto Verganti, Harvard Business Press, Boston, MA, 2009.
7　*Winning at New Products: Creating Value Through Innovation* (5th ed.), Robert G. Cooper, Basic Books, New York, 2023.
8　CapEx (Capital Expenditure) refers to funds used by a company to acquire, upgrade, or maintain physical assets such as buildings, machinery, or equipment. These are long-term investments intended to improve or expand operations.

7 Understand (Get It)

CONSUMER CENTRICITY

The first phase of the 8-step process is the Understand phase. Food and beverage innovation becomes truly impactful when it revolves around the consumer and the full context of their life. Understanding the consumer and the different situations they encounter in daily life is essential to subsequently understanding what they would like to buy. There is a very thought-provoking movie clip on YouTube with Noel Gallagher,[1] talking about understanding the consumer and why it doesn't make sense to ask them what they want. It is fun and worth watching, so if it is still to be found, please do so. He argues that groundbreaking art isn't shaped by consumer demand, but by artists taking bold risks. He points out that the world didn't ask for revolutionary artists like Jimi Hendrix or iconic albums like Sgt. Pepper's, but once they arrived, they reshaped the culture.

According to him, the music industry has become too focused on catering to consumers and playing it safe, which stifles innovation. He compares this to fashion, where designers don't rely on focus groups or consumer feedback to decide what's next. Instead, they push boundaries and dictate trends, driving change forward. Gallagher believes that letting consumer preferences dictate creativity limits progress, because most people don't truly know what they want until they're shown something new and extraordinary. Yes! That is a very bold statement, and to some extent, indeed, consumers don't always know what they want. However, in music, art, and fashion, the motivation is often not to create a product for the consumers. Motivation is a creative desire to express yourself as an artist, and if other people like it – great! Rick Rubin, producer behind many successful bands, as Beastie Boys, Red Hot Chili Peppers, Run-D.M.C., Metallica, Johnny Cash, et cetera, wrote a book, titled "The Creative Act: A Way of Being."[2] In this book and in several interviews, he describes many of the creative aspects of making music and puts emphasis on the importance of not having a consumer in mind when composing music. The examples from Noel Gallagher are certainly in this category, and art in general should be.

These examples are also examples of success where the songs of the artist resonated with a larger audience and hence became a viable business. Think of all the artists composing and publishing their art each day, and how many of them actually

DOI: 10.1201/9781003619352-9
This chapter was refined for grammar and fluency using ChatGPT-5.0.

succeed in turning the art into a source of income that can pay their bills and provide food on the table.

The food and beverage industry, however, is not art! It can be creative and fun, yes, but at the end of the day, the business is about producing products that the consumers would love to buy and consume. To increase the probability of success, it is essential to understand the consumers who will be your target audience. Who are they, and what is their situation? Who will buy and consume the product, and why? I think Andrew Geoghegan puts it very well in his book, *Effective Brand Building*[3] – from a brand perspective.

> ... – *We ask ourselves how our brand can take advantage of what we are seeing, not about how our brands can add value to people. To be human-centric marketers we need to put people first.*

The same goes for product innovation! We need to put the consumers first! A classic way to explore this is by asking fundamental "W-questions" and a How?

1. Who
2. Why
3. When
4. With
5. What
6. Where
7. How

These questions create a logical consumer-centric flow by firstly defining the audience, then understanding their motivations, context, product choices, interaction points, and their behavior. **Let me elaborate:**

Who
Who is the target audience? This includes where in the world they live, their lifestyle, habits, preferences, and values that influence their purchasing and consumption behavior. As mentioned, it is important to distinguish between the person making the purchase and the person consuming the product, as their motivations may differ.

Why
Why would a consumer choose to buy this product? This question explores the motivation behind the purchase. Is it driven by functional or emotional needs, personal preferences, social influences, or constraints? In all cases, it will be a mix of many factors. What does the consumer need or desire, and why will a product fulfill this? Does it solve a problem, enhance an experience, offer something new, provide added convenience, quality, or value? – and so on. Understanding what drives demand, helps to understand the opportunity and shape a relevant and compelling product.

When
When does the consumer need this product? Timing plays a crucial role in consumer behavior. Is it associated with a particular time of day, season, or occasion? Occasions

during a day can vary widely, and could be highly individual, depending on the lifestyle and life situations.

With

This considers the social context of consumption. With whom the consumer are when enjoying the product, is important. Are they alone, with family, colleagues, or friends? The social setting can influence purchasing decisions and consumption habits.

What

What do consumers currently buy and use? This question focuses on existing choices and consumption habits. Understanding what products consumers already purchase provides insights into market trends, preferences, and potential gaps. It also helps to identify whether consumers are brand-loyal, price-sensitive, or open to new types of products in their food and beverage choices.

Where

Where do consumers interact with the product? Where do they discover, purchase, and consume it?

- **Discover:** Is the first encounter with the product by recommendations from family or friends, via social media, advertisement, in-store displays, et cetera?
- **Purchase:** Where is the product bought? In a supermarket, a convenience store, a specialty shop, through an online platform, et cetera?
- **Consume:** Where is the product typically enjoyed? At home, on the go, in the workplace, in another social setting, et cetera?

How

How will the consumer use this product? Is it consumed as is, combined with other products, or integrated into a specific routine? Understanding how the product fits into daily life provides valuable insights for the physical product design, the design of the packaging, and how to communicate what the product is about.

Exploring these 3 behaviorisms is very valuable:

- **Consumption Style:** Is it eaten or drunk immediately, prepared in a specific way, used as an ingredient in a larger meal, et cetera?
- **Routine Integration:** Does the product become part of a morning ritual, post-workout recovery, social gathering, a quick snack on the go, et cetera?
- **Purchase Behavior:** Are consumers planning their purchase, adding it to a regular shopping list, or is it typically bought on impulse due to cravings, promotions, convenience, et cetera?

The order of the questions can vary a great deal from project to project. It all depends on your starting point, and as mentioned earlier, it might be very different. If you start with a very broad pre-scope, the consumer segment could be the point to begin to understand what unmet needs they have. If it is with a higher focus on an occasion, investigate the consumer moment first. If you already have certain constraints on category and product type, or even a great idea – start to understand that space first – and be pragmatic! The timing of the innovation project is also a key consideration to include. Consumer preferences change over time, and anticipating these changes helps to ensure that the launching of a product is relevant to a significant audience at launch and remains to be long enough to achieve a sensible return on investment. In the chapters to come, we will uncover all the questions, using different methodologies and tools.

Enabling Innovation by Empathy

Empathy is the ability to understand the feelings, thoughts, and perspectives of others. In a consumer context, it involves being able to emotionally resonate with the tensions in daily life. Understand what motivates them and what is holding them back. What drives their purchase intention? I will state that empathy is the backbone of innovation. For me, innovation is when creativity, enabled by empathy, is successful.

"Innovation is when creativity, enabled by EMPATHY, is successful."

If you don't understand the consumers that you are trying to invent new food and beverage products for, there is a high risk that the product will be unsuccessful. We will get back to the part of creativity later. Let's dive deep into the ocean of empathy. How does it translate into a food and beverage context?

An Example of Cooking for a Friend:

Imagine you are cooking a meal for a friend and want to make her a dish she loves. You might recall previous meals you have shared and think about what she enjoyed the most. Did your friend talk about a particular dish she enjoyed? Little things count. Did she avoid something on her plate? Maybe you know she loves a sprinkle of extra cheese, or she has a favorite herb, loves lemon, et cetera. This attention to detailed memories offers insights into the likes and dislikes of your friend. Use empathy as well to put yourself in your friend's shoes. If you know she had a stressful week, maybe she would appreciate comfort food. If she has recently been through a healthy phase, perhaps she would prefer something green, light, and nutritious. Your friend's cultural background can furthermore deeply influence her food preferences, and you need to consider that as well.

At its core, cooking a meal your friend with love is about more than just the food. It's a gesture of care, affection, and appreciation. Taking the time to truly understand her taste preferences and current emotional situation is a way of saying, "I value our relationship."

It is a bit the same when you are aiming to design a new retail food or beverage product. To creatively innovate solutions that truly make a difference for consumers, you must deeply understand for whom you are solving a problem, and why it is a problem.

INSIGHTS HARVESTING AND PLANNING

When starting a project, it is always advisable to review the knowledge already available within the company, especially if it is in some way relevant to the pre-scope. If the material comes from a previous project, some perspectives might not fully align with the new area of focus. Be mindful of this but focus on identifying the insights that still hold value, and harvest them. Large companies will most likely have a lot in their archives from previous projects, and it can be tempting to conclude that there is too much material to go through. Before the help of AI was possible, this task could actually be a major barrier, but not anymore. More importantly, it could be that some of the data is no longer valid, due to the age of the material. It could be that it is 10 years old, and the world and consumers have changed significantly. It was a different target audience, another category, and so on. Legit arguments, if that is the case, and it is for sure something to be aware of; however, it is also important to know where the consumers are coming from. How have the shifts in behavior been? Who were the first movers in the past, and are they actually more normal in the present environment?

Out-of-scope or dated insights can still be valuable and helpful when planning what to investigate further. Smaller companies, or new companies, might not have that luxury and will have to start from scratch. In the coming subchapters, I will list a lot of different types of research that you can initiate to become wiser in the area of the pre-scope. Obviously, a single project will not be able to initiate everything listed. It would be way too expensive, but also not valuable. Some types of research will not be relevant or just "nice to have." The art is to pinpoint what the "need to knows" are, and how to acquire that knowledge in the best possible way. The cheapest way can sometimes be the best way. The fastest way, can sometimes be the best way. The premium quality way, can sometimes be the best way. Use the intelligence within your company to assess this, but also take advice from outside experts, if needed.

There are no guarantees of success in our business, but you can acquire data and translate it into insights that make the risk of failure less likely. Let's start with how to understand more about the consumers and their lives. Their attitudes and beliefs, social and cultural influences, emotional needs, aspirations and goals, behaviors and habits, taste and dietary preferences, cooking and food shopping habits, time management, decision-making processes, financial situations, health and wellness priorities, environmental and ethical concerns, sustainability preferences, dreams and inspirations, challenges or pain points, and all the connections between them. Yes – it is a lot!

CONSUMER RESEARCH

Most small- and medium-sized food and beverage companies do not have their own consumer science researchers. If that is the case, my clear recommendation is to reach out to supporting companies that provide this. Defining the target audience to fit a pre-scope can be difficult, and it is the entire foundation of all the work lying ahead.

Use consumer science expertise to guide this.

It is not always just about identifying the core group of people that potentially could fit the "end consumer" of what the innovation project will eventually result in but also identifying the consumers who are on the outskirts of the target audience or even very different types of consumers. They can provide perspective on the behavior of the target audience and might even illuminate unmet needs and potential demand.

SEGMENTATION

The world is a vast ocean of different types of consumers, and they can be segmented in many ways. The classic approach is to view them through four categories of segmentation, used in different combinations.[4]

1. Demographic Segmentation: this involves dividing the market based on variables such as age, gender, family size, income, education, occupation, and nationality.
2. Geographic Segmentation: this type separates consumers based on their physical location, such as country, region, city, or neighborhood.
3. Psychographic Segmentation: this approach looks at psychological traits of consumers, including lifestyles, values, attitudes, and personality characteristics.
4. Behavioral Segmentation: this type focuses on how consumers act concerning the product, including knowledge of, attitude towards, use of, or response to a product, service, or brand.

Depending on the target of the research you wish to conduct, you will typically screen consumers by looking at the four categories and selecting specific variables relevant to what you want to investigate. If you, as an example, want to create a segmentation of individuals with busy careers, but also family responsibilities, use the four categories as a guide to what is important for the scope of what you want to research. It could look like the following:

Demographic Segmentation:

*   Age: Typically, in the 30–50 age range.
*   Gender: All genders.
*   Family Size: Usually includes children, often ranging from 1–3 kids.
*   Income: Middle to upper-middle income, often dual-income households.
*   Education: Generally well educated, with at least a bachelor's degree, often with professional or advanced degrees.
*   Occupation: Professionals in demanding careers such as law, finance, healthcare, education, or corporate roles.

Geographic Segmentation:

*   Market: The country in play.
*   Location: Medium to large cities. Might live outside the city and commute.

Psychographic Segmentation:

- Lifestyle: Busy, focused on career, with significant family responsibilities. They often juggle work and home life, striving for work-life balance.
- Values: Highly value efficiency, productivity, and the well-being of their family. Health and nutrition are important, but convenience is a necessity.
- Attitudes: Practical and results-oriented. They prioritize finding solutions that help them manage their time effectively while maintaining quality of life.
- Personality: Typically organized, responsible, and driven, with a strong desire for achievement and fulfillment both professionally and personally.

Behavioral Segmentation:

- Usage Rate: Frequent users of convenience products and services, especially those that save time in daily routines.
- Benefits Sought: Efficiency, ease of use, and products/services that support a balanced lifestyle. Conscious of health, but time-constrained, seeking quick and healthy options.
- Loyalty Status: Often loyal to brands that consistently deliver on convenience and quality, with a preference for solutions that reduce daily stress.
- Occasions: Situations with time pressure, such as rushed mornings, quick lunch breaks, and evenings with limited time for meal preparation.

Based on the consumer segmentation scope, questions can be formulated to screen the consumers relevant to inform the pre-scope. Certain categories could have a higher focus, for example the behavioral segmentation, including sub points like the occasions and sought benefits.

New Dawn Example:

To whom would it make sense for **The Golden Juice Company** to research?
    It could be something like this:

Demographic:

- Age: 30 - 50 years age range.
- Gender: All.
- Family Size: 1 - 3 kids.
- Education: Generally well-educated, with at least a bachelor's degree.

Geographic:

- Market: United Kingdom, Germany, The Netherlands.
- Location: Medium to large cities. Might live outside the city and commute.

Psychographic:

- Lifestyle: Busy, career-focused, to some degree, but with significant family responsibilities. They often juggle work and home life, striving for a work-life balance.

Behavioral:

- Benefits Sought: Efficiency, easy use, and products/services that support a balanced lifestyle. Health-conscious, but time-constrained, seeking quick and healthy options.
- Occasions: Situations with time pressure, such as rushed mornings, quick lunch breaks.

The segmentation can also be illustrated using more tangible personas. However, this step should follow after the consumer research, as the insights gathered will provide factual data from real lives to create them. More on personas and how to build them will be explained in the Chapter on the Define Phase .

First-movers or Last-movers

Another lens is to look at when and to what degree the segment adopts innovations on the market. Everett Rogers'[5] Theory of Diffusion of Innovations, classify consumers based on how soon they adopt a new idea or product after its introduction. These categories are useful for understanding the adoption process within various groups and can help in strategizing the rollout of innovations. He divides consumers into five categories and also has an estimate of how big a part of the population it is:

1. Innovators (2.5% of the population)
This small category of consumers is also widely known as *first-movers*. They seek and explore new launches in various categories. Within the context of food, they will often try out new restaurants and follow other first-movers on social media to be up to date on what to try next. They will also embrace new launches of food and beverage products as the first. These curious consumers are very valuable to talk to in the early phases of an innovation project. They can help to assess and provide feedback towards a more radical innovation of products, concepts, or even new categories. They are way too small an audience to become the target audience for most companies, but they can help to guide the direction towards potential roadmaps of innovation.

2. Early Adopters (13.5%)
Early adopters are the next group to try a product launch. They will for sure follow the first-movers and get inspired by them. The early adopters are the ones who give a new food trend traction. They are important to talk to in the iterative process of testing and adjusting prototypes and concepts.

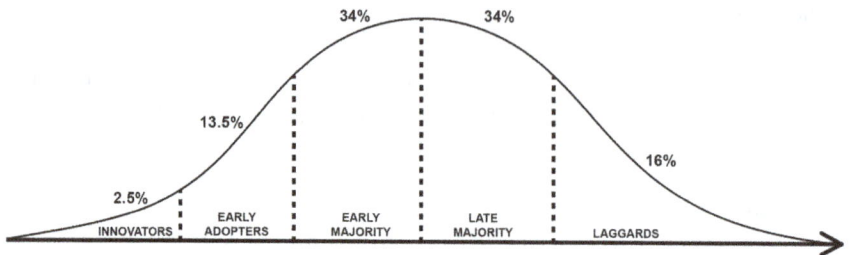

FIGURE 7.1   Diffusion of Innovations Theory Bell Curve (Re-visualized by the author based on the work of Everett Rogers.)

3. Early Majority (34%)

The early majority adopts new ideas just before the average member of society. They need evidence that the new products work before committing. They are also important to talk to in the iterative process of testing, and furthermore, they can guide you on the strategic timings of a potential launch. They can also give indication of volume in a concept test.

4. Late Majority (34%)

The late majority are skeptical of change and will only adopt an innovation after it has been on the market for quite some time. They should be considered in calculating the growth potential. If you reach them, your product is mainstream.

5. Laggards (16%)

Laggards are the last to adopt an innovation. If you reach them, your product is a classic.

With a good idea of how to segment the target audience, and which other interesting consumers could be helpful in your project, you need to decide what to ask them and how to ask them, or how and where to observe them.

TYPES OF RESEARCH

Research can be divided into two overarching types:

1. Qualitative research (Qual)
2. Quantitative research (Quant)

QUALITATIVE RESEARCH

Qualitative research is a collection of exploratory methods used to gain deep, contextual insights into consumer behaviors, motivations, and attitudes. It explores "the why" behind how people think, feel, and act. It is a key enabler to understand and uncover consumer needs. Frequently, there are needs they can not articulate, or even might not be aware of. The needs can be met with current products, fully or to some extent, or even be unmet needs. If real unmet needs are identified, they hold immense innovation potential. Qualitative research is based on collected data that cannot be measured traditionally, such as assigning it a numerical value. It could, for example, be an interview with a target audience where you discuss how they live their lives and maybe what morning routines they have. The data gathered from an interview are the statements provided in the conversations. Their experiences and opinions. The result will be an assessment by the interviewer and a translation of what the statements could mean. Qualitative research is also mainly characterized by a lower number of participants. Ranging from around 4 - 6 participants to 40 - 60. All depends on the type of research.

Interview with a Qualitative Research Expert:

During the research for this book, I have had several qualitative interviews and conversations around the topic of design thinking applied in food and beverage innovation. Around the topic of research, I had an interesting interview with a very experienced consumer researcher, sharing perspectives on the value of qualitative research. Pernille Holm Vind is the founder of a Danish company,[6] specialized specifically in qual research. For years, they have been mapping the food habits of the Danish consumers. Pernille finds the qualitative methods especially valuable to understand different perspectives on consumers' lives. They explore these perspectives together with their customers to provide them innovation directions. During our conversation, Pernille emphasized that a key approach to qualitative research and segmentation, for her company, is maintaining a strong focus on consumers' life phases, particularly during times of significant life changes. She illustrated this with the mindset shift around food habits that often occur when a young couple is expecting their first baby. Before the pregnancy, they might not have given much thought to their choices of food, but with the approaching change, they start to evaluate how they have nurtured their bodies and how they aspire to in their future family. In a way, you can say that the transition from a certain life phase into another amplifies the reflections and makes it easier to verbalize in a conversation. I believe this is a valuable perspective to keep in mind when engaging with consumers. Even though Pernille is a huge advocate of qualitative research and its value to understand the deep emotional layers of consumers and their needs, she also acknowledges the value of the other types of research, e.g. the quantitative research.

QUANTITATIVE RESEARCH

Quantitative research is a systematic method used to measure and analyze consumer behaviors, preferences, and trends using numerical data. It answers the *what* by showing how many people think, feel, or act in specific ways. Quantitative research is based on collected data that can be put in numbers. It could be an online survey with a tailor-made questionnaire, asking a target audience different questions regarding attitudes and behavior. Many times, they will consist of multiple-choice questions that, in the result, can be quantified. It could also be rating a statement on a scale to what extent the respondent agrees or disagrees. Quantitative research is mainly characterized by a higher number of participants. Ranging from around 100 participants to thousands. It all depends on the type of research.

Use a Combination of Qual and Quant

In much of the design thinking literature, quantitative research is not regarded as highly as qualitative research. However, the best approach is to use both. While qualitative research focuses on understanding the *why* behind consumer behavior and attitudes, quantitative research helps determine how prevalent they are. The sequence in which they are conducted should align with what you're trying to achieve. If you're exploring a very broad scope, start with qual to understand the area more deeply. If you need statistical validation, start with quant and follow up with qual for depth.

But also, a more iterative approach mixing both methods from the start can be valuable.

What is important to consider when designing consumer research?

The way you frame questions and conduct the research significantly impacts the quality of the insights. Basically, you want to ask consumers in ways that don't guide them too much towards specific answers that you might wish for. In qual, it is important to use open-ended questions to encourage storytelling about real-life behavior and attitudes rather than simple yes/no answers. In quant, it is important to design structured questions with clear response options to ensure measurable and comparable data, allowing for statistical analysis of patterns and trends.

The Say – Do Gap

In all types of research, there can be a gap between what consumers say they do and what they actually do. It is easier to identify in qual research, as you can ask follow-up questions if needed. People in general normally try to show the best version of themselves and their lives when asked. When, for example, you ask consumers how often they buy organic food in a quant survey, it will most likely be far higher than the actual sales figures. They might have the wish and intent, but for multiple reasons, it might not translate into an actual purchase. It can be that the price is too high, or the availability of organic products is low, where they shop. It might be that they don't have an actual intent, but answer positively as they follow the social norms of what is appreciated behavior in their local society.

Consumer research in the Understand phase and what to initiate depends on the questions raised in the pre-scope of your project. But let's take a broad approach and explore some relevant research options to consider. I will exemplify with the New Dawn innovation project, but it is not to say that it will be required to initiate all of them to gather the needed insights.

N.B. I will always recommend consulting with consumer science professionals to choose the right approach for your innovation project.

Focus Group (Qual)

This is a concept Noel from Oasis hates! But actually, it is a very valuable means of research to get a deeper understanding of the *why*. You can have many types of focus group setups, but basically, it is inviting a group of people to have a facilitated discussion on a topic. Depending on what questions you need answers to, you invite people within a relevant target audience. For example, a certain Juice company from UK is exploring an innovation project with a wish to produce RTD smoothies as the solution! Let's say they also have an assumption that this solution could be aimed at busy professionals starting their day, skipping breakfast, with the consequence of running low on energy. With all that pre-scoped, they could arrange a focus group with that target audience to explore how they integrate or would wish to integrate energy and nutrition into their on-the-go morning routines.

The discussion could focus on why some people choose to skip breakfast entirely and the potential consequences of doing so, as well as the alternatives they choose. Whether it is grabbing something *on the go* or eating at the office. If none of the participants mentions smoothies as an option, the interviewer can ask why that might be and explore smoothies as a potential choice: why some don't prefer them, what alternatives they choose instead, and the reasons behind those choices. The discussion should be broad and explore the entire occasion, the consumers' motivations, as well as the functional and emotional needs. The interviewer will prepare some different inputs to show the group and curate a discussion going through a well-prepared interview guide.

It takes a lot of knowledge and experience to be a good facilitator of an open discussion. It is very important not to steer the conversation too much and ask closed questions. It is also important to have the session in an unbiased setting and not reveal who you are as a brand. The main reason to have focus groups is to gain a deeper understanding of attitudes and needs. What are the tensions in their lives, and what drives the motivation for them to act on those needs and tensions? What are the barriers they face?

If relevant to the desired outcome, you can reveal the brand at the end of a session and have an open talk about how the participants view the brand's *right to play* in the smoothie category. The focus group session will be recorded, and afterward, you can analyze the statements and start to create some learnings and new assumptions.

Online Surveys (Quant)

This is where you can measure stuff! In the digital world of today, it is very easy to get consumers to answer online questionnaires. Many supporting companies offer this type of service, and they also have huge consumer audience databases to segment and ask. They will offer both self-service solutions and customized research by their experts. The questions will typically fall into several categories, depending on what you are aiming for. It might be that you want to ask questions to measure behaviors, attitudes, and preferences around smoothies. New assumptions that you might have put together based on a focus group session! Assuming they align with the target audience from the qualitative research and are at least open to using smoothies, whether or not they currently consume them. Here are a few examples of how questions can be phrased and what possible answers they can choose:

Behavior:

- How often do you make your own homemade smoothie?
  (Daily, Weekly, Monthly, Rarely, Never)
- How often do you purchase Ready-to-Drink (RTD) smoothies?
  (Daily, Weekly, Monthly, Rarely, Never)
- How often do you consume smoothies as a meal replacement?
  (Daily, Weekly, Monthly, Rarely, Never)
- At what time of day do you typically consume smoothies?
  (Morning, Afternoon, Evening, No set time)
- Where do you usually purchase your RTD smoothies?
  (Supermarket, Convenience store, Coffee shop, Online, Other)

Attitude:

- How effective do you find smoothies as a light meal option?
  (Not at all effective, Slightly, Moderately, Very, Extremely effective)
- How effective are smoothies in helping you reach your daily fruit and vegetable intake?
  (Not at all effective, Slightly, Moderately, Very, Extremely effective)
- How important is nutritional value when choosing a smoothie?
  (Not important, Slightly, Moderately, Very, Extremely important)
- What is the most important factor when choosing an RTD smoothie?
  (Taste, Ingredients, Price, Brand, Convenience, Other)

Preference:

- Which of the following ingredients do you prefer in your smoothies?
  (Check all that apply: Banana, Berries, Spinach, Protein powder, Oats, Nuts, Other)
- Which type of smoothie do you prefer?
  (Fruit-based, Vegetable-based, A combination of Fruit and Veg, Other)
- How sweet do you prefer your RTD smoothies?
  (Not sweet at all, Lightly sweetened, Moderately sweet, Very sweet)

Based on the data collected, you will be able to get a better assessment of how prevalent the behavior, attitude, and preferences are towards smoothies during the morning. If you decide to use the self-service surveys, there are some rules to be aware of. To write good survey questions, keep them clear, simple, and neutral. Avoid words or terms that are too professional. Think about whether an average consumer would know a term like that. If needed, explain what the term is about. Also, avoid leading questions and make sure the questions are relevant and only ask what matters. It takes time for the participants to fill out these questionnaires, and if it gets too complicated, illogical, or even boring, they will click through it fast without the necessary reflection, or simply stop answering. Offer balanced answer choices, including neutral options like "Not sure," when needed. Structure the survey logically, starting broadly and getting more specific.

Ethnographic – Field Study (Qual)

A more analog approach is the field study. If you want to enter a new market or consumer target audience where you really have a gap in the fundamental understanding of the behavior and motivations, being "a fly on the wall" can be very beneficial. This type of research is especially relevant in mid- and long-term projects as it is a time-consuming type of research and the learnings need to be reflected upon. More about this topic in the chapter, "Future foresight." In qual fieldwork like this, it can be challenging to create an environment where the observer does not influence the behavior of the participants. For instance, when visiting a family to observe their daily routines and dietary habits, your presence might unintentionally prompt them to prepare healthier meals than they normally would. To mitigate this effect, I recommend involving skilled anthropologists to conduct this type of research.

Consumer Video Diaries (Qual)

Another online option for the field study is to have consumers record short videos, allowing you to observe their real-life behaviors and decision-making processes. The participants will need a guide detailing the specific tasks and questions to address. These recordings can capture moments such as the following:

Shopping experiences:
Following them as they navigate stores, make purchasing decisions, and explain why they choose certain products over others.

Pantry and fridge insights:
A peek inside their fridge and cupboards, revealing what they stock, how they organize it, and the reasoning behind their purchases.

Daily routines:
Understanding how they incorporate specific products into their daily lives. From morning rituals and snacking habits to lunch, dinner, and everything in between.

Web Communities (Qual + Quant)

This approach is to facilitate discussions in an online community setup, allowing for both qualitative and quantitative insights. Many companies provide platforms where larger groups of participants can engage in discussions on specific topics, either in real time or at their own pace. This setup allows researchers to observe key areas of interest while also engaging directly with participants, encouraging them to elaborate further.

Artificial Intelligence and Research

Generative AI might shift these traditional research methodologies, such as surveys and focus groups, toward more observational, data-driven approaches. But then again, we need to remember that the data needs to come from humans! As of 2025, the field is advancing rapidly, and I anticipate that AI will have a paradigm-shifting impact on consumer research in the coming years. I am hearing about AI persona bots that can answer questions and potentially engage in conversations about a segment's motivations and behaviors – helping to identify unmet needs and foster consumer empathy among participants in ideation workshops. It truly sounds amazing! It will help to create better ideas that solve real consumer problems. I also see that insight companies take AI beyond research, and have it create the ideas as well … Ultimately, we wouldn't need humans. AI could create every-thing from insights to products! Seems too science fiction! – right? But one thing is certain. AI is here to stay, and the companies that embrace the technology and learn how to use it in the best possible way will have an edge over competitors that ignore it! AI's ability to analyze vast datasets can reveal hidden patterns in con-sumer preferences, behavior shifts, and emerging trends. This is for sure faster – and maybe even better!

Social Media Listening

Social media listening is the process of monitoring social media platforms to track conversations that consumers have. AI has transformed this field from a primarily qualitative approach, where humans manually investigated platforms and selected interesting statements and conversations, into a more automated, data-driven process. Now, large datasets of social media content are available online, which can be effi-ciently processed and analyzed by AI to uncover insights. Several insight companies offer social media listening solutions.

How can this data be used?

Using demographic data, social media listening can help segment the audience and identify personas. For instance, you could analyze which age groups or regions are most interested in specific topics. This is part of the segmentation exercise. Automated qualitative research, where AI can derive insights from natural conversations, reviews, and even visual content shared by consumers, can be very useful in predicting trends. Beyond identifying current trends, social media listening data can also be used to pre-dict future consumer behavior. Using social media listening is a very powerful tool to identify both behavioral trends and food and beverage trends.

These trendy conversations are interesting to understand and translate into public sentiment insights. Public sentiment, often referred to as public opinion, is the col-lective attitude or opinion of the population of a society or community on specific issues, trends, or topics. It is also a useful way to get insights on competitor per-formance, as we will look more closely at in the chapter, "Market and Competitor Analysis."

WHAT TO SELECT AND INITIATE?

These different types of consumer research come in many varieties. Each offers unique insights to create value in your innovation project. Choosing the right approach depends on several factors connected to the project pre-scope. Some methods, like online surveys and web communities, are great for gathering broad quantitative and qualitative data, while video ethnography and in-depth interviews provide richer, more contextual insights into consumer behaviors and motivations. By selecting the right combination of methods, you can uncover meaningful patterns, validate assumptions, and make more informed decisions on what to take forward in your innovation project. This will provide the needed understanding and empathy to ideate against later in the process. Based on the insights from what research is initiated and already available in the "company drawer," let's enter the work of further exploring the pre-scope and mapping out essential narratives about the target audience, their lives, and the F&B market dynamics. As previously touched upon about enabling innovation by empathy, let's dive deeper into understanding what jobs consumers are hiring food and beverage products to do.

JOBS TO BE DONE-THEORY

I really like the Jobs to Be Done-theory[7] (JTBD), introduced by Clayton M. Christensen and his co-authors. It provides the analogy that customers *hire* products or services to accomplish specific tasks or goals in their lives. To explain the concept of the product JBTD analogy, I will start with an example from outside the food and beverage consumption world.

Imagine that you have a transportation problem to solve! Let's say you need to go from your home to a party at your friend's house.

Taxi Example

You could bike, but your home is quite far from your friend's house, and it would be exhausting to bike, and sweat on arrival. You could also take public transportation, but you are not fond of crowded places with a lot of strange people. Furthermore, the trip will take a long time. The taxi is more expensive than the two other options, and you are considering paying the extra cost to get the convenience, driving comfort, and speed. You have previously hired a taxi … but it is expensive! Suddenly, you remember a friend telling you about a new app-based driving service, which should be convenient and a fast way to book a car ride. You download the app, register, and 30 minutes later, you arrive at your friend's house, in a private car, at a much lower cost compared to a taxi. In essence, you have previously *hired* different solutions to do the job of transportation for you, and in this case, you decide to hire a new one. It is a bit the same in the food and beverage world!

Mayo Example

As an example, you might *hire* a squeeze-bottle of mayonnaise from your local supermarket to solve a problem around making a delicious sandwich for your evening meal. You already have the bread, cheese, ham, lettuce, tomato, onion, mustard, and cornichons, but you need that final ingredient, the dressing, to spread on the sandwich and complete the taste experience you cherish. You are a bit of a foodie and actually capable of making your own homemade mayonnaise. However, sometimes you get a result that tastes a bit weird, or the texture is too thin, and it also takes up more time. This is what we call a barrier to fulfilling your desire for homemade mayo in your sandwich on a typical day! In this case, you decide to *hire* the bottle of mayonnaise you know has a good taste and texture. The motivation is the convenience of not having to mix a homemade version with the uncertainty of success, the taste, functionality, and speed. This way of looking at ordinary situations in consumers' lives and identifying the jobs they *hire* products to do can reveal what they are truly trying to accomplish. Some motivations are obvious and may seem too basic to articulate. Buzzwords like *convenience*, *price*, and *taste* will always surface, and that is perfectly normal. The trick is to look deeper and understand why convenience is important. What compromises are consumers willing to accept to achieve it? How do they use creativity to make inconvenient products work in certain situations?

Look for all kinds of details and translate them into important things to consider when innovating new solutions. New solutions that do the job better than what consumers hire today, or provide them with solutions they didn't even realize they needed. According to Christensen & Co., it is recommended to use the data from both consumer research and competitor analysis, but it is also perfectly valuable to use your personal observations. Look at your own life and how you interact with food and beverages. It is highly likely that if you hire a product to do a job, other consumers will do the same. Trust the value of your own and your colleagues' points of view.

N.B. This is actually more complicated in business situations than you would imagine. Especially in larger companies, it seems like personal observations and opinions are not valid. Even though I am all for data and the objectivity of research, I am also a strong advocate for using intelligence within company employees. Therefore, a big kudos to the authors for emphasizing this. Furthermore, they pinpoint areas to help identify JTDBs by looking at consumer behavior, such as when they have a need, but choose not to solve it with existing products. The reason could be that they don't know what could solve the need for them at all, or the solutions available have too significant side effects to be attractive.

Help to Sleep Example

For example, many consumers struggle to fall asleep and seek natural solutions to help them. Many avoid sleeping pills for various reasons, such as the fear of addiction or social stigma if family and friends find out. Instead, they might prefer a natural food or drink option to help them relax and feel sleepy. They have heard the old advice about drinking chamomile tea, warm milk, or even tart cherry juice before bed, but these options come with an unfortunate downside, as they often lead to waking up in the middle of the night to use the bathroom. They might even try to solve the problem with workarounds, such as reducing screen time, reading before bed, or relying on the old trick of counting sheep. The job to be done for an F&B product would be the instant gratification of getting sleepy, without a side effect that will wake you up again.

Lastly, it is also very interesting to identify when consumers use certain products in unusual ways. Those types of workarounds could hold great innovation potential.

BREADCRUMBS EXAMPLE

Breadcrumbs are a great example of how consumers really get creative and use all kinds of products to coat their meat, fish, or veggies. Have you ever tried a schnitzel coated in TUC crackers? It gives a crunchy texture and a combination of sweetness and saltiness; or salmon coated in a mix of crushed wasabi peas and Panko?; or deep-fried broccoli with crushed cup noodles in the batter? Look for funny taste hacks like this, or more subtle alternative uses of familiar products in cooking and consumption behavior. The breadcrumbs examples obviously hold inspiration, if you are in the breadcrumb business, or a ready-meal producer, but also try to look at it more broadly.

The choices and experiences could inspire innovation within many categories. How would these hacks translate into beverages or desserts? Would unfamiliar ingredients or products provide a new taste or functionality in an ice cream? Have you tried pistachio ice cream drizzled with virgin olive oil and topped with a little sea salt? It's delicious! The JTBD theory covers all types of products that people use, and I want to highlight the importance of taste when looking for JTBDs in the food and beverage industry. It is typically the most important aspect of the product experience for consumers, so consider it when looking for the jobs to be done. To some consumers, taste might actually not be the most important aspect of the JTBD, and that is also very important to capture if that should be the case. Much more follows on the importance of taste in the chapter: "Taste is king."

WHAT IS HIRED TODAY?

Later in the process, we will use the JTBD framework to analyze specific consumer segments and identify problems that F&B products can solve, unlocking innovation potential through a deeper understanding of consumers' lives. To challenge conventional thinking, food and beverage consumers aren't really looking for mayonnaise, spreads, or dressings in the supermarket. They're looking for something to *hire* to add creaminess and taste to their sandwich, and to meals in general. In this way, JTBD thinking challenges the notion that consumers are merely buying products. Instead, they are *hiring* solutions to accomplish a specific job in their lives. We will also explore what consumers use to fulfill needs by examining the products they currently rely on to get the job done. It could be the case that some of the products they hire today are very good solutions! Some products might be very similar to what your company is able to produce. Try to identify why these solutions are chosen, and why do they solve the job?

Also highlight any barriers that can be identified, such as being unhealthy, messy, or expensive. Look at the barriers and think of them as opportunities to solve for the consumer. This analysis can uncover gaps or opportunities for improvement. In summary, identifying the jobs to be done involves asking and observing consumers to understand, not only what they choose to do, but also what they choose not to do, due to emotional, social, or functional barriers, and lastly, how important the taste experience is to them. **Some of the barriers** might be the mess in the kitchen, cooking skill frustrations, or other obstacles that are the friction points to uncover the **unmet needs** of the consumers. I am not sure how the innovation process behind the first factory-made mayonnaise was conducted. The insights and mindset behind it, but I am sure that for the first consumers trying it, they discovered an unmet need that they might not even have been aware of. Unmet needs may not be immediately visible or easy to spot, but developing this skill is essential for creating opportunities for innovation that can lead to targeted solutions. More on that topic a bit later, but before that, let's look into what happens before consumers even decide to buy a product.

What makes consumers decide to try a certain product, and what needs to be fulfilled for the purchase to actually happen? That is also an essential topic to understand. JTBD is a valuable framework for understanding consumer behavior, but it is more complex than that. Consumers aren't simply shopping to solve abstract needs without having a clear preference, and they know exactly which products and brands they normally use or want to try. How do the expectations build, and how do they end up assessing the product experience? An experience that will influence what they do next. Let's map out all the moments before, during, and after a purchase. A consumer-purchase-journey-map!

THE CONSUMER PURCHASE JOURNEY

When a consumer decides to buy a food or beverage product, they are influenced by a variety of input. All the input provided from different sources accumulates into a set of expectations for a product. It is part of the consumer journey. An essential input that the consumer will have is the product pack design. With a deep understanding of the target audience and drivers of purchase, you can determine how to compose the pack design and create the product expectations that will resonate with them. When the consumers bought the product, they will consume the product and enjoy the product experience. If their expectations are met or exceeded, they will most likely rebuy the product. In the book *ID*,[8] David Airey frames it very well, stating:

> ... *people are smarter to keep buying something that sets expectations higher than what's delivered.*

With a deep understanding of the target audience and drivers of product preference, you can determine how to design the product recipe and create the product experience that will resonate with them. Basically, it can be framed into six phases[9] where consumers are influenced by a huge amount of input and emotional and functional needs.

1. Awareness
2. Consideration
3. Purchase
4. Experience
5. Re-buy
6. Advocacy

During the first four phases, all the **expectations** for the product are built, while in phase four, the **experience** starts. Let's investigate what it could look like:

1. AWARENESS
Awareness is the first phase of the consumer purchase journey. This is the point in time where the potential consumer is made aware of a product for the first time. It does not actually have to be a new product on the market, but rather new to this specific consumer. Let's call her Carrie. Carrie is influenced by many inputs from outside, but also by her own feelings, needs, and desires.

Emotional and Functional Needs:
Carrie's personality reflects the sum of her life experiences and the predispositions of her genetic origins. What she finds attractive is closely connected to her values and emotions. In line with these she prioritizes certain properties of the food and beverages she consumes. It might be physical or mental health-related properties she seeks. Or she could also care less about that and be more the indulgent type:

Who cares if it is healthy I only want to enjoy life and have a delicious treat when I feel like it!

Carrie could also be the type that views food only as fuel to provide energy and satiety. It is not a constant state, and Carrie's emotional and functional priorities will most likely change during her life. One of the reasons for this is that Carrie is not alone in the world.

She is influenced by the people around her!

Family, Friends, and a potential Partner:
The influence of friends and family plays a significant role in the awareness phase. What is most important, it all depends on your personality and stage of life. For some, what their best friend says carries more weight than the statements of their father. And vice versa. Or in some cases, the choices and statements from people not that close to you, but people that you look to, get inspired by, mean more. In Carrie's case, she might secretly be in love with someone in her network and, as a result, pay especially close attention to what that person considers the "next big thing."

Media:
What the next big thing could be is again influenced by what is buzzing in society. It can be trends around food and beverages but also shifts in behavior and attitudes against certain topics. Carrie will notice these movements in many ways. Relations as one source. Advertisements and social media as another. Carrie uses several social media platforms and gets inspired, influenced, and aware of new product offerings and new cool food and beverage hacks to try out. It will be both influencers she follows, but also all the advertisements that follow, and are included in the apps. She might still be watching Flow TV, with commercial blocks, or she has switched to the more personalized streaming services and the complimentary targeted ads.

If she also subscribes to lifestyle magazines, she will also see selected reviews of products, even though that part is to some extent also being taken over by social media influencers. But how this split will be in the future is hard to predict. With all the information just described, Carrie's awareness of certain topics or products will settle in her consciousness, and she will carry it with her into the next phase. The phase where she starts to consider whether the new functional smoothie with added collagen, her best friend mentioned, is worth trying. A seed of expectation for a potential future experience has been planted.

2. CONSIDERATION

Carrie is maybe not consciously thinking about buying any product yet, but she starts to think more about it and notices more information on the topic or product. At some point, the consideration will materialize and become conscious. And this is where it takes off! Most adults know the strange feeling of suddenly noticing the car or bike they are considering buying, seemingly everywhere they go. On the road, in parking lots, and even in advertisements, as if the whole universe is conspiring to grab their attention. This is the same thing! Expectations are growing when confirmed that others also like it enough to buy it. If it is a specific product, like the functional smoothie with collagen, and it is available in Carrie's local supermarket, she will most likely notice it while shopping. She might take a glance at the price and move on the first day she spots it, but she might also step into the next phase. The actual purchase phase.

3. PURCHASE

Carrie is in the supermarket again! Expectations and considerations are set in motion, and she is ready to act on her impulse. She wants to try the smoothie. Will she be able to find the product again? It could be sold out, or simply not available at the specific supermarket. (Crucial factors to be in place, if we want Carrie to buy our product!)

Other crucial factors also worth mentioning are the following:

- Will the smoothie stand out on the shelf in a way that makes it different from the other smoothies on the market?
- Will she notice the one she heard about, or does a competitor attract her attention more effectively?
- Maybe the visual appeal and colors of the competition are spot on Carrie's preferences. The brand identity plays an important role in Carrie's choice.
- Maybe the visual cues on the competitor's pack, signaling how delicious the taste and texture are, are more appealing and fit Carrie's taste preferences better.
- Maybe the functional claims and other emotional health cues will skew Carrie's emotional judgment towards the alternative.
- Will the packaging format, quality, and size matter to her?
- Will the competition have a better price, and will it matter to Carrie?

Design thinking can help you make the right choices, assessing how to design and communicate, and if the competition doesn't exceed your capabilities, you will have Carrie in the palm of your hand.

All the impressions Carrie takes in, within just a few seconds, build her expectations even further. In this example, she ends up buying the functional smoothie she was recommended. Happy about the choice she does her additional shopping and heads home. She is looking forward to her morning commute to work the next day, where she intends to have the smoothie as a breakfast on the go. Carrie is heading into Phase 4: The product experience.

Phase 4 is essential, as it is the total collection of product features that will influence her consumer experience of the product. It is where it will be revealed whether the expectations built in the first three phases are met or even exceeded by the product experience.

4. EXPERIENCE

The next morning, Carrie heads to the office with the chilled smoothie in her bag. At the metro, she drinks the smoothie alongside all the other morning commuters, and the experience is influenced by many organoleptic impressions. Before she opens the bottle, she looks at the design on the pack. The brand logo, the colors, and the text about the product composition, the claims, and the taste cues. Carrie chose the green smoothie with the text:

Green Glow Smoothie, packed with all the good stuff to keep you feeling as bright as your day deserves to be! We blended:

Bananas for sweetness and smoothness.
Spinach because greens are great.
A splash of **Lemon juice** for a zesty zing.
Kefir for some gut-loving goodness.
A pinch of **Spirulina.**
And a dash of **Collagen** to keep you glowing from the inside out

Totally natural, 100% tasty, and made with love (and absolutely no funny business).

The messages on the pack make Carrie feel healthy, and her expectations of a fresh and delicious experience, are increasing. She also notices the tactile sensation of the bottle material, and it feels like good quality which enhances her experience of having bought a premium product!

She shakes the bottle and listens to the sound, and observes the product color and the viscosity through the transparent bottle. Hearing the familiar click of the screw cap, she twists it open and catches the first scent of sweet fruitiness from the banana and the freshness from the lemon. Carrie likes the very pleasant sensation and takes her first sip. Carrie might not be a sensory expert and wouldn't be able to describe the multitude of sensations she is now experiencing, but here is what it could look like:

The texture and mouthfeel:
The texture of the smoothie is thick, as Carrie expected from a smoothie. The mouthfeel is smooth and creamy, with no particles to be detected.

The basic taste balance:
There is a good balance between the sweetness from the banana, the acidity from the kefir and the lemon juice – a sweet and sour taste. There is also a slight bitterness from the spinach and spirulina, but not too overpowering.

The flavors from the different ingredients:

The dominant flavor is the banana. A very familiar and recognizable flavor. The spinach also provides a subtle, unique green veggie flavor to the product. The kefir brings some fermented dairy flavor to the product, and there is a small touch of lemon flavor. The spirulina (a cyanobacteria, often called blue-green algae) brings a slight seaweed sensation as well.

The taste experience is liked by Carrie, very much! The expectations were not only met – in this case, the experience was even better than Carrie had imagined.

The total experience is very positive! As Carrie uses the smoothie as her breakfast, she will also think about how full it made her feel, however, not completely full as a first sensation.

She turns the bottle and takes a look at the nutritional information to see how much protein or fibre it actually contains. It contained: 1.6 g protein and 0.7g fibre, pr 100g! Not that much to get full … maybe I should bring a protein bar as well, she reflects. With this nearly perfect new product experience in Carrie's mind, she enters the last phases of the consumer purchase journey. The re-buy and advocacy phases.

5. RE-BUY

The re-buy phase is where the scores are calculated, and the consumer evaluates how well the expectations for the product were met in the experience. If there is a good match, the consumer might re-buy the product, and if it is a "spot on innovation," the product will become a frequently bought item. The consumer will most likely also start to tell friends and family about this amazing product, and the process will start over for all the new potential consumers in the awareness phase. In Carrie's case, she went back to the supermarket and decided to **re-buy** more of the smoothie. The range came in 6 different flavor variations, and she tried some of the others as well. She also bought some Oat and Chocolate protein bars to provide her with more fullness for her "on the go" breakfast occasion.

6. ADVOCACY

When she met up with her friend who made the recommendation, she talked about the experience and thanked her for the great tip. Later on, she also started to tell other friends and her family about her new breakfast setup with positive statements, as follows:

It is really tasty for such a healthy product!

Did you notice the new glow and vibrance of my skin, ha ha? :-)

I feel it is a little expensive, but definitely worth it.

You can sometimes buy it on offer, and if you combine it with these protein bars, you will be full until lunch, for sure.

Carrie has become an **advocate** for the product. Based on her positive experience and attitude, she is likely to help the brand and the product succeed.

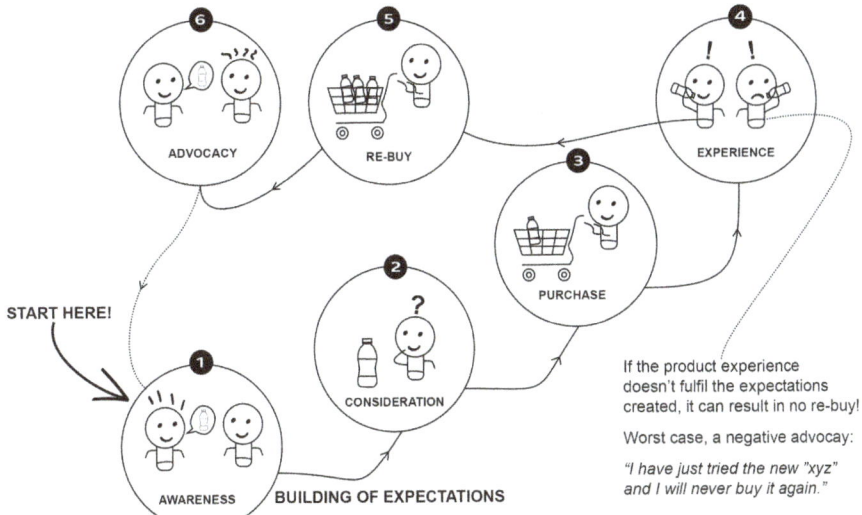

FIGURE 7.2   The Consumer Journey.

As indicated in the illustration, Figure 7.2, the phases are tied together and will run in circular loops. The consumer's purchase journey is crucial to understand:

Why and how are consumers influenced to try new products?

How are the expectations built, and how is the experience perceived?

By putting that knowledge together with an understanding of what kind of job the product is hired to do, and what it is hired today, will create a strong foundation to a design thinking process.

OCCASIONS

In food and beverage innovation, we often talk about *occasions*. It is typically referring to when and where the moment of consumption takes place. But what truly shapes these moments isn't just the time and place of the event. It's the entire context and the underlying motivation behind why the moments take place. In everyday language, we often ask, "What's the occasion?" This is not just to mark a point in time, but rather to understand why something is happening. In that sense, *occasion* can also mean the motivation, or the reason behind a particular act of consumption. In this book, an occasion refers to both the context and the underlying motivation for a moment when food or beverage consumption takes place, or could potentially occur.

BIRTHDAY EXAMPLE

You decide to bring a cake to the office to share during a break. When your colleagues see you with the cake, they will most likely ask:

What's the occasion?

They already understand the context, as they are at the office with you, during a shared break. What they're really asking is: Why did you bring a cake? What was the motivation? Your answers could reflect a range of motivations. Some are obvious, others are more implicit and often left unspoken. You might answer the following:

It is my birthday, so I wanted to celebrate that with you!

The true motivation could also be: It is an unwritten rule to bring cake to the office on your birthday, and you have no desire to break that.

Or you might answer:

I just love to bake and made too much!

But the real motivation could be a desire to be liked at the office, or to show some skills.

UNDERSTANDING DAILY PATTERNS

To build deeper understanding and empathy with a target audience, mapping a typical day in their lives can help to identify where meals, snacks, and beverages are, or could be, consumed. Most importantly, however, is why? The illustration, Figure 7.3, shows what a day could look like for an average consumer working at an office. There can be small differences from day to day. Maybe Monday and Wednesday are the Gym days and that will change the pattern compared to the rest of the weekdays, and so on.

Throughout the day, consumers move through different patterns that can reveal both met and unmet needs, as well as motivations, attitudes, and behaviors related to food and beverage consumption. Of course, no two days or people are exactly the same. That is why it is important to talk to consumers about what a "typical day" looks like for them. Ask about the context of their consumption moments, as well as the motivations behind them.

Context Question Examples

Questions to help you understand the situation and environment around each moment of consumption can be the following:

- What did you consume?
- Where did the consumption take place?
- What was happening at that time? (e.g., work break, relaxing, commuting, et cetera.)
- Who were you with, if anyone?
- How much time did you have?
- Were there kitchen facilities nearby? (e.g., a microwave, fridge, kettle, et cetera.)
- Was there space to sit down and eat or drink comfortably?
- What types of stores or food options were accessible nearby?

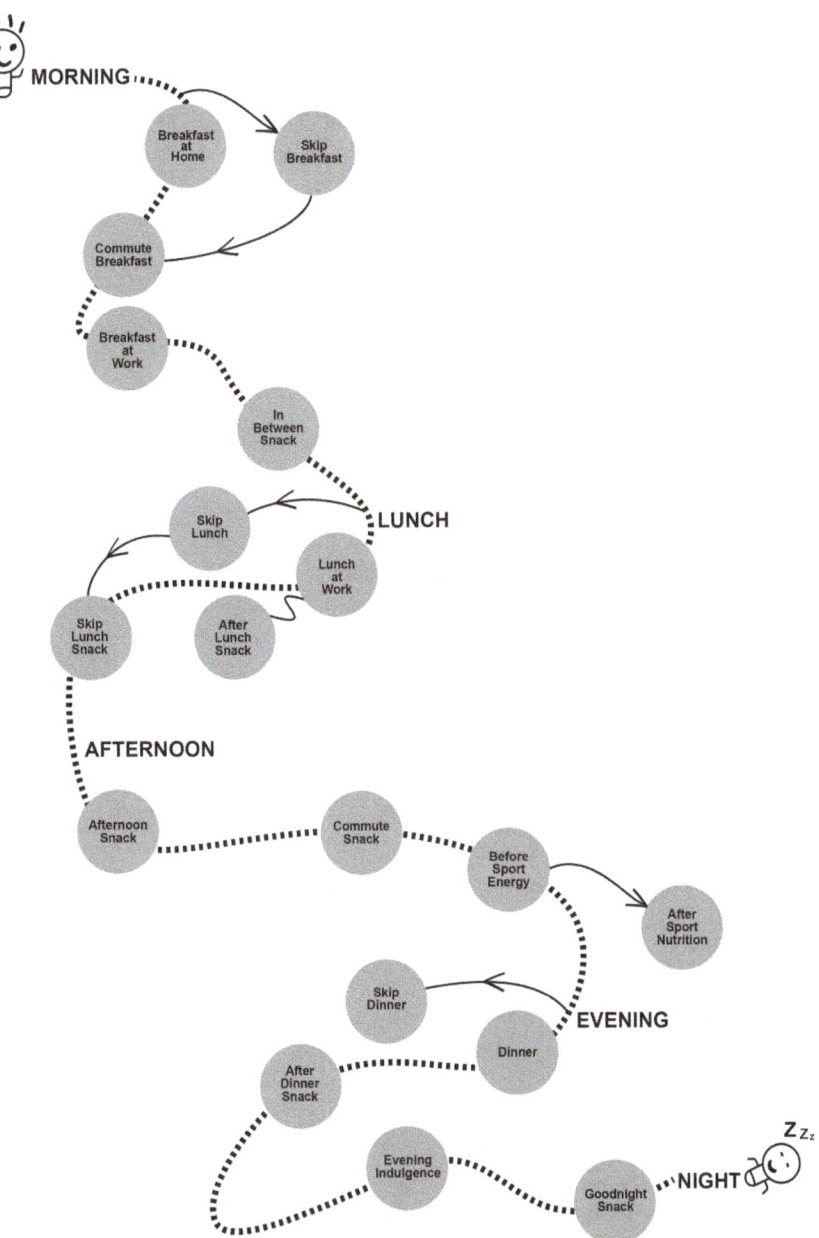

FIGURE 7.3   A Day of Choices.

Examples of Motivation Questions

Be mindful that motivation questions only make sense when asked in relation to the specific context of the consumption moment. Understanding the motivation always depends on the situation. The same person might choose very different things in the morning rush, during a relaxed evening, or after a workout.

Here are some examples of questions to help you uncover motivations:

- What made you decide to eat or drink something at that moment? (e.g., indulgence, refreshment, hunger, low energy, et cetera.)
- Did you plan ahead for this moment, or was it spontaneous?
- Did you skip a meal earlier? If so, what influenced that decision, and how did it affect this moment?
- Were you craving something specific? If so, what and why?
- Did anyone or anything influence the moment – such as the product you chose, the place it happened, the timing, or how much time you had?

The questions can be asked during several of the types of consumer research we just went through earlier in this chapter. The answers will provide you with different descriptions of real-life occasions that can help to frame a potential demand and opportunity related to the project scope.

For now, let's visit Carrie again and imagine how a typical day could look for her from an occasional point of view.

It is possible to split the day in many ways, but a simple approach could be the following:

Morning: 5:00 AM – 8:00 AM
Mid-morning: 8:00 AM – 12:00 PM
Lunch: 12:00 PM – 2:00 PM
Afternoon: 2:00 PM – 4:00 PM
Late Afternoon: 4:00 PM – 6:00 PM
Early Evening: 6:00 PM – 8:00 PM
Evening: 8:00 PM – 10:00 PM
Late Evening: 10:00 PM – 12:00 AM
Night: 12:00 AM – 5:00 AM

The time ranges, of course, can vary based on cultural, regional, or individual preferences. These fit with Carrie's life.

I recommend writing a consumer occasion journey like a story. It makes it more engaging and easier to empathize with her situation. If possible, it would also be valuable to have a sketch artist illustrate the day as a comic strip to have a quick visual overview. It will make the story even more relatable and make it easier to go through and discuss as a team.

The Story Could Be Something Like This:

A Monday morning at the outskirts of London, UK.

...

The sound of the alarm pulls Carrie from sleep at **5:30 AM**. She stretches briefly in bed, enjoying the quiet moments before the day's chaos begins. Sliding out of bed, she drinks the glass of water on her night table and stretches her body. As the kids start to wake up, her partner is showering. She heads to the kitchen and refills her glass with her herbal tea. She realizes that she could really use something more filling, but with little time to prepare anything for herself, she pushes the thought aside and heads for the bathroom too.

At **6:00 AM,** the house is alive with activity. She and her partner prepare breakfast for the kids and their lunch boxes together – both simultaneously checking the day's emails on their phones. Carrie wonders if there's a better way to start her mornings, but the clock keeps ticking.

At **6:45 AM,** they are on their way to drop off the children at school, which luckily is located near their apartment.

By **7:30 AM**, Carrie is on her commute on the metro. The coffee thermo cup provides her with a caffeine kick as she listens to a podcast about time management. The thought of her missing breakfast lingers, and she makes a mental note to pick up something more substantial later in the day. The office is already buzzing when Carrie arrives at **8:30 AM.** Back-to-back meetings fill her calendar, one blending into the next.

At **9:30 AM**, she grabs a quick coffee from the office pantry to keep her energy up. She notices the cookies in the snack basket but hesitates and ends up not taking one. While it's a nice treat, she wishes there were healthier options to get her through the morning without a sugar crash.

By **11:00 AM**, she is starting to become really hungry, and her energy level and mental clarity are going down. Carrie has no time to leave her desk, as she needs to reply to something before lunch. She remembers the cookies from earlier, not ideal, but it will have to do.

At **12:30 PM**, she finally has her lunch break. Carrie picks up a salad from a supermarket nearby, brings it back to her desk, and eats it quickly, while scrolling through news updates and emails. She thinks about how repetitive and boring her lunch routine has become and wishes she were more up for a social talk with colleagues while having lunch. The afternoon is a whirlwind of calls and deadlines. At **3:00 PM**, Carrie heads back to the pantry for more coffee. This time, she grabs yet another cookie without hesitation. By **4:30 PM**, Carrie is wrapping up her day. The ride home feels long, and she nearly falls asleep. Today her partner will pick up the children, but she needs to do the shopping for the evening dinner and what else they need for the week before returning home.

Back at the apartment at **6:30 PM**, her focus shifts to talking with the kids and her partner and preparing dinner.

Tonight, it is a mix of premade salads, hummus, falafels, and bread. Quick, healthy, and satisfying. She wonders, though, if there's a way to get even more time to prepare dinner, like her mother did. She would love to cook more from scratch and give that meal experience to her children.

By **9:00 PM**, the kids are finally in bed, and Carrie sinks into the couch with a cup of chamomile tea, together with her partner. She considers having a small dessert, but decides not to. Mostly out of guilt for the cookies earlier. Instead, she grabs a book about corporate strategy and lets herself "unwind" for the first time all day.

At **10:30 PM**, Carrie prepares for bed, reflecting on the busy and productive day. As she sips water before turning off the lights, she makes a mental note to find better snacks to bring during the commute and for the office. Something her best friend told her over the weekend pops into her mind! Tomorrow is another day, but she's determined to make it smoother …

It is possible to build the narrative and map the journey in many ways. It doesn't have to be the entire day, either, if the project scope is within a certain category that prompts for certain parts of the day. It can also be short and sweet, in bullet points like the following:

- **Location and time:** Commutes to work, 7:30 AM.
- **Action:** Drinks coffee from her thermos cup.
- **Occasion:** On the go "breakfast".
- **JTBD:** Provide energy and mental sharpness.
- **What's hired today?** Coffee.
- **Barrier:** Time pressure to prepare something more substantial to bring and consume.
- **Unmet Need:** The sensation of a meal and the feeling of being full.
- **Motivation:** The desire to bring more health and quality into her life.

In the consumer occasion journey, there are a lot of barriers that, in some consumers' minds, create tension, but in others don't. It is an area of the consumer's life that is really important to deeply understand .

In Carrie's case, she hires a coffee to do the job as an "on the go" breakfast. But does it do the job perfectly, or is there an unmet need that could improve the experience even better?

She heard about a smoothie from her friend that might be a better hire. It actually might be that she would prefer not to buy a finished smoothie and prepare it herself based on her wish to prepare food from scratch. But certain barriers make her buy the "ready to drink" products after all. Barriers could be time in the morning. When she needs to commute to work, that takes time: time from her sleep and kids, which she values. She would also need to buy a range of ingredients to be able to prepare the smoothie. That takes time, too. Even if she decided that the time barrier would not prevent her from making her own morning smoothies, other barriers would be likely to emerge. Would Carrie be able to compose a delicious smoothie with the taste and texture she appreciates? Would she hate the mess and having to clean the blender? Annoying work and yet another time constraint, she didn't consider. Barriers and unmet needs can be many, but sometimes hard to put your finger on them. Many are hidden in plain sight. She is certainly motivated to look for a better alternative, but not to the degree that homemade will be something she is likely to try.

Spotting barriers and unmet needs and assessing them against the underlying motivations that could trigger a behavior change, is an essential discipline in design thinking for F&B innovation.

Some situations involving food and beverage consumption have been part of people's lives for decades, and they might not even realize these could be seen as problems that would make life easier if addressed. But if a new solution is introduced that solves the problem and makes the situation much nicer and more convenient, suddenly they will reflect on how annoying it was previously. Have you ever thought about how annoying the convenient teabag is?

THE TEABAG EXAMPLE

In the world of tea today, there are many ways of preparing your tea, but one of the most widespread ways is the teabag. The teabag was invented around 1900,[10] to address the need to brew single servings without loose leaves floating around in the cup. A real problem was solved, but a problem that many tea drinkers before the introduction probably didn't think about. Later in time, the bag format also created the possibility to display different types of tea in one box, which has become a huge hit in offices and restaurants. Giving both variety and convenience.

Are there any downsides to the teabag? Certainly! The tea quality is not always the best, and you can certainly make more flavorful tea using whole leaves and more traditional preparation methods. However, that takes more time and effort. Therefore, depending on the consumption moment, some consumers prioritize that. In other instances, the time and place are not suitable and the teabag is convenient.

Could it be more convenient? Yes! Because of the decades of usage, the inconvenience of the teabag is probably not top of mind to consumers, but there are still some barriers that potentially could be solved. When you disperse your teabag into the hot water, it will take a little while before the flavor of the tea has been extracted into the water. Assuming you are working at an office, during the small waiting time, you might move away from the kitchen and back to your desk. The tea extraction is done, and you would like to drink your tea. What to do with the teabag? In this example, you have some options, but all are a bit inconvenient:

1. You can leave your teabag in the tea, with the consequence of getting bitter notes in your tea and an unpleasant drinking experience.
2. You can take the teabag out and place it on something like another cup or plate if that is within reach, with the consequence of making a mess on your desk and more into the dishwasher.
3. You can go straight back to the kitchen or the nearest garbage bin and dispose of the teabag.

The average teabag user most likely don't think much about the small inconvenience of all three options, but if it was solved with a new solution that didn't remove any of the good things about the teabag, but just provided you with the tea of your choice in the same way as the coffee machine at the office provides you with coffee, without handling beans or filters, it would make your life easier. A barrier, thus, can be top of the mind for a consumer, or not even noticed. Being able to identify those barriers in the lives of consumers is a crucial building block in understanding their unmet needs.

The Attached Screw Cap Example:

A "barrier" can also work the other way around. Sometimes, a new solution is introduced not because of an expressed consumer need, but because of a broader agenda from society. In these cases, consumers may not initially welcome the change, especially if it impacts their routines or feels less convenient. An example is the recent EU legislation mandating attached caps on plastic beverage bottles. The aim is clear: to reduce plastic litter by ensuring that caps are not discarded separately. By attaching the cap to the bottle, it is deemed more likely to be recycled together with the bottle, reducing environmental harm. From a functionality standpoint, however, this change comes with friction.

For years, consumers have twisted off bottle caps and tossed them in the bin – or unfortunately, sometimes on the ground. With the new attached design, the cap hangs awkwardly while you drink, often getting in the way, hitting your nose, or even causing spills on your clothes. It is reasonable to assume that some consumers find this new drinking experience frustrating.

To wrap up the two examples, the good news is:

Whether a friction is hidden or obvious, it always holds innovation potential!

Past, Present and Future Changes

It is within human nature to evolve, and evolution can be measured by looking back at what has happened. Looking at what is taking place in the present, and the difficult one, trying to predict what is going to happen. The design thinking for food and beverage innovation needs to take all three points in time into consideration and should understand these changes. As previously discussed, there are different kinds of research you can initiate to understand the present situation of the consumers – the consumers that you are aiming to develop new products for.

Research should be done to understand the jobs to be done by these products, and also how consumers live their lives. A life can be broken down into a vast set of short moments of consumption. A life can also be influenced by a multitude of factors in the purchase journey of shopping for the nutrition needed to both sustain life, but also to make it both more effective and enjoyable. In the coming chapters, there will be more detail on how to investigate the overarching trends in consumer behavior, the market and competitor movements, product trends, and future foresight, and why they all are important factors in designing food and beverage products.

Consumer behavior spans across all three points in time, whereas market and competitor analysis are mostly looking at the past and present. The same goes for identifying product trends, but they can also transcend into future predictions on how a behavioral trend, materialized in a new type of product, could evolve. Future foresighting as a discipline draws insights from the past and the present to anticipate scenarios that the food and beverage industry must prepare for.

BEHAVIORAL TRENDS

Behavioral trends often reflect evolving consumer priorities, as well as shifts in attitudes and habits. Understanding these trends and the fundamental human needs behind them is crucial for identifying opportunities and gaining the empathy needed to innovate successful products or services.

Which essential human needs are important when you want to understand the underlying factors behind consumers' food and beverage choices?
Humans reflect complex connections between the mind and the body, a dynamic interplay that shapes physical and psychological performance, health, and well-being. Human behavior and needs are several fields of science put together! It is not the aim to cover all aspects and details of human physiology and psychology, but let's take a closer look at some of the basic needs and activities, and how they interlink with food and beverage choices.

Physical Needs: Proper nutrition is essential for sustaining life and overall well-being. The physical needs include macro- and micro-nutrients, water, exercise, and rest.

Mental and Emotional Needs: These include cognitive functions like thinking, reasoning, memory, focus, and problem-solving, as part of the practical and logical mind. The other part is **emotions**. Emotional well-being involves having self-esteem, the ability to express emotions, and having meaningful relationships, all of which are essential for a good life. Together, these needs shape how individuals think, feel, and respond to life's challenges.

FUNCTIONAL NEEDS

Activities that support **physical** needs include:

- Eating balanced meals.
- Staying hydrated.
- Incorporating regular cardio and muscle-strengthening activities into a weekly routine.
- Maintaining a consistent sleep schedule.

Activities that support specific **mental** needs, like memory, focus, and reducing stress, can be the following:

- Different types of mind exercises.
- Meditation or mindfulness.
- Incorporating foods rich in certain ingredients and micronutrients into the diet, which are known to improve or maintain the specific area.

EMOTIONAL NEEDS

The most powerful drivers behind food and beverage choices are often emotional. Some examples of emotional needs that are related to food and beverages could be:

- **Social Needs:** Sharing meals with family, friends, or colleagues to foster meaningful relationships and an emotional connection – these are a fundamental part of life.
- **Pleasure Needs:** Focus on enjoying sensory experiences, like appreciating the textures and flavors of a meal or beverage. Indulging in desserts or comfort foods, and trying new cuisines, or other culinary experiences, is for some people a huge part of life.
- **Self-expression Needs:** Food and beverage choices can reflect an individual's personality and values. These can be shared in social interactions or on social media. Showing and posting products, or dishes, can be a way of letting people know how cool you are.
- **Value Needs:** These reflect the principles that influence consumer decisions beyond practical or economic considerations. For example, this could include consumers who prioritize living more sustainably by choosing food with a low CO_2 impact, locally sourced, or seasonal ingredients, and habits that reduce food waste.

Based on the pre-scope, investigate the relevant consumer behavior. This type of research can be a combination of quantitative and qualitative methods, as described earlier. Reports on behavioral trends are also available from specialized research companies, using trend sociologists. These companies often use different descriptors and methodologies, typically organizing their findings into overarching themes called *Megatrends*. Beneath these, they highlight shorter-term shifts and patterns referred to as micro trends, which are often illustrated by using a wheel with the consumer at its center.

Megatrend research looks at long-term shifts that influence human behavior across society, including the economy, culture, technology, and the environment. These broader changes not only shape what consumers buy, but also how they connect their food choices to their values and lifestyles.

Microtrends are more niche, local, or short-term shifts in consumer behavior, attitudes, and usage. Unlike mega trends, which span across years and have a broad societal impact, micro trends tend to emerge quickly, have a smaller reach, and often appeal to specific subcultures or market segments. Whether a trend is truly mega or still micro can, of course, be up for debate, as some trends sit in a gray area. It all depends on how widespread and long-lasting they have become. Trend analysis, in general, is often very focused on specific products, which is both logical and valuable. However, it is worth the effort to understand them in relation to behavioral trends, as well as their interconnections. We will deep-dive more into the analysis of the food and beverage trends later. Firstly two examples of current megatrends within the food and beverages which have been prevalent for years and most likely will continue to be, looking ahead.

Health and Wellness Trend

This megatrend centers on what benefits the consumer's personal health. It is driven by a combination of emotional needs, such as the desire for well-being and taking care of yourself, and the functional needs, like nutrition, fitness, and disease prevention.

With a heightened focus on personal health and longevity, many consumers seek to change their behavior. They may start to exercise and try out functional food and beverages to support the body's recovery after exercise. Products are those with added functionality and benefits to support mental wellness, gut health, immunity, energy levels, and so on ...

There are four basic attitudes towards personal health:

1. **Prevent:** Avoid harm and anticipate risks (e.g., healthy eating, regular exercise).
2. **Maintain:** Keep the current state of health or wellbeing (e.g., routine check-ups, consistent habits).
3. **Improve:** Actively enhance physical, mental, or social health (e.g., learning new skills, therapy, or fitness goals).
4. **Neglect:** Take no action, whether due to lack of awareness, resources, motivation, or prioritization (e.g., ignoring symptoms, delaying decisions).

The barrier lies in the reasons why people neglect their own health and well-being, while the solution lies in how food and beverage products can support prevention, maintenance, and improvement. Understanding these behavioral patterns better can enable F&B companies to address deeper motivations and design meaningful solutions for health and wellness.

Sustainable Diet Trends

This megatrend focuses on what is beneficial for the planet's health and how food and beverage consumption impacts it. It is primarily driven by an emotional need to make ethical and environmentally responsible choices.

Sustainable food and beverage choices are often seen as a single concept, but they can address different environmental challenges. Three key behavioral shifts are influencing the consumers:

1. **Lowering Carbon Footprints:**
   Lowering carbon footprints is about reducing greenhouse gas emissions from food production and transportation.
2. **Minimizing Pollution:**
   Food production contributes to various forms of pollution, including water contamination, soil degradation, and plastic waste. Pesticides, fertilizers, and antibiotics used in industrial farming contaminate the ecosystems, while single-use packaging can lead to plastic pollution.

3. **Reducing Food Waste:**
   Optimizing food production and consumption to prevent loss and waste.

While all three help to prevent environmental damage and optimize the use of planetary resources, consumers are often confused about the distinctions, assuming that any sustainable food choice automatically addresses all issues. However, the reality is more nuanced. Because all three topics are part of the sustainability discussion, they often get mixed up in consumer perception, subsequently leading to oversimplified conclusions, such as the following:

- Eating locally is always better for the planet, even though imported lentils often have a lower carbon footprint than locally produced beef.
- Buying organic always means it's more sustainable, although it may still have a high carbon footprint.
- Conventional food has higher yields and is more efficient, yet it can still harm the environment through soil degradation from intensive farming and pollution from fertilizers and pesticides.

Regardless of the complexity, sustainable food and beverage choices are some of the behavior changes that many consumers feel is both relatable and actionable. It is easy to implement and to communicate to family, friends, and colleagues.

I am trying to make a difference by eating more plant-based food, I always buy organic, and I have started to use all the leftovers.

Behavioral trends are not equally important all over the world. The first two examples provided are the most common in developed countries, while in less developed regions, people tend to prioritize sustaining life over these lifestyle trends. The last example is a combination of emotional and functional needs, with a high emphasis on the functional need of sustaining life.

Affordable Nutrition Trend

Across many developing and emerging markets, affordable nutrition is a dominant megatrend, driven by the intersection of economic constraints and nutritional needs. While the specific food solutions vary by region, the underlying emotional and functional needs shaping behavior remain consistent.

Emotional Needs: Food security and predictability

- In many households, the fear of food scarcity shapes the purchasing behavior, prioritizing reliable, but affordable staples over variety.
- Food is of high emotional importance, representing care, survival, and dignity.

Functional Needs: Maximizing nutrition within a budget

- With financial constraints, food choices are made based on cost-effectiveness per calorie and nutrient, favoring filling, and energy-dense foods.
- Affordable sources of fortified foods and essential nutrients are increasingly in demand, particularly for children and vulnerable parts of populations.

Behavioral Trend: Frequent purchases and trusted brands

- Instead of bulk buying, many consumers rely on daily or frequent small purchases, adapting to inconsistent income flows.
- Consumers gravitate toward brands they trust, particularly those that balance affordability with perceived health benefits.

Affordable Nutrition is not just about cost it is also about stability, trust, and maintaining cultural and social dignity while adapting to financial realities.

The behavior and life situations of consumers are a constant change and adaptation process. The behavioral trends are reflected in what consumers choose to buy and how they use the current products available to them. In that way, there is a circular dynamic of what consumers have a need for, and care about, and what the food and beverage industry decides to launch and provide them. In the following chapter, we will look more at all the tangible F&B outcomes of behavioral trends, but of course also other factors – as what the companies would like to produce for various reasons. We are zooming in on what is currently available on the market.

- What are the competitors up to?
- What is really selling well and being talked about?

What was available, but didn't make it?

MARKET AND COMPETITOR ANALYSIS

THE MARKET SITUATION

In the food and beverage industry, whatever category it might be, you will have competitors in the market. Most likely, many of them! The barriers to innovate, develop, produce, and get a listing are relatively low, allowing many players to introduce new products, which makes it a constant challenge to stand out. Market saturation often leads to a need for innovation, better branding, and products or price differentiation to capture the consumer's attention. Gaining consumer loyalty is particularly difficult! Price-conscious consumers frequently switch to other brands or a private label when they perceive better value, making consistent brand loyalty rare.

By comparison, the car industry operates at the opposite end of the spectrum. High costs to enter a market and long development cycles mean far fewer competitors on the market. However, once established, car manufacturers often enjoy longer product life cycles and greater customer loyalty compared to the F&B sector. Not every country has its own car manufacturers, but all countries have their own food and beverage producers.

In the F&B industry, it can be established companies that dominate a specific market, having been there for years and years, or new players trying to enter the category with a new offering. This is true, especially if we look at local markets in isolation. Every country has its own local F&B industry and its products are often strong players in terms of hitting the local taste preferences. They are good at getting distribution due to the power of a local network. Some of the companies are also present in other markets, and they have to know these markets as well as their own. The closer a company gets to global distribution, the more difficult it becomes to know each market's small differences. But they still need to try and learn from what the local companies are doing.

In the same way, smaller and more local companies will need to look at what bigger and more global companies launch and have success with. Global companies will naturally also compare themselves with other global players. Here, the number of players drops significantly. The interesting question, regardless of company size or market dominance, is why the competition, big or small, has success with a product? In other words, why are consumers buying it?

The first question to investigate is what new product launches have entered the market in the last 2 - 3 years? The next very intriguing question is, whether the competition is successful. One thing is to be overwhelmed by crafty products, powerful design, and the USPs of a key account manager's dreams coming true, but does it sell? Try to get sales data to assess the success of certain products of interest. Choose the best performing ones and deep-dive into questions like the following:

- What problem does the competitor solve for the target audience?
- What is the taste profile of the product and of the other competitors?
- What is the price of the product, and what is the price level in general of that type of product? Compare the unit price vs. the kg price.
- What *on-pack claims* are used? And what other design elements does the packaging consist of?
- What are the ingredients used? How do the other products on the market compare with these products?
- What is the nutritional content? And how does it compare with the other products on the market?
- What are the key messages and strategies used in the competitors' marketing communications for the product?

Certainly, there may be products analyzed without being proven successful. Perhaps these products may still be new on the market and a step ahead of you. Other launches might have been too early or failed due to many other reasons. One may also look at other markets of interest as well – markets that have a similar profile or the opposite. Very different markets can provide inspiration and perspective. To identify which market is the leader within the product categories of your company is essential! So, how to analyse it? Let's look at ways to conduct market and competitor analysis.

THE ANALYSIS

An effective market and competitor analysis is like a puzzle, composed of various pieces of information. To gather the information and analyze it, take time and resources, but it is a valuable input in designing the product features later on in the process. Once again, it all depends on the number of resources your company can invest in the area.

To get a good overview, there are several areas to investigate:

- Sales data
- New launches data
- Online store check
- Company website communication
- Advertisement
- Store monitoring
- Consumer shops along
- Social Media listening
- Product reviews and articles

Pick the ones that are relevant to answer questions related to the pre-scope of your project. It might still be a question what product category will be the intended outcome – and many times, not! All the areas above can provide very valuable insights to add to the full competitor analysis. In addition, I also recommend doing market and competitor analysis proactively, and not related to a specific innovation project. The more you know about what is going on in the product categories core to your company, the better you will be able to assess and scope innovation potential within the category.

Sales Data

Whether in any way, it is possible to obtain sales data from a relevant category and market, it is the best starting point for a valuable analysis. This may be an important piece of the puzzle! Before using too many resources on the details of a competitor's product design, it is really crucial to know if the competitor's product is actually successful. To assess the potential of a new product or concept, it is very advisable to get sales data on similar products and markets. The classic way is to subscribe to sales data within a core category and market. Or, buy specific bespoke reports. Several companies offer this type of service. The data can, however, be expensive.

As mentioned earlier, retailers have a big advantage here, as they have sales data from their own organization. This is a big advantage in private label innovation projects. As a F&B producer or brand, it is sometimes possible to collaborate with retailers and access some of their sales data. This is mostly the case in a close partnership.

New Launches Data

When working within F&B innovation, all new product launches are interesting to see and to taste. Innovators will always keep an eye out for new products on the market. Regardless of whether the pre-scope includes a product or a category constraint, you need to acquire knowledge about the product categories your company already produces and adjacent categories that it could feasibly expand to. To get the big picture, it can be necessary to subscribe to a service that monitors all the new launches of products. There are once again several companies offering this type of service. What is launched is not showing how successful the product is, and to my knowledge, it is still in 2025 not possible to subscribe to a service that combines new launches with actual sales data. I am sure, though, that it is just a matter of time before a combination will see the light of day. Some newly launched services use AI to scan the online availability and that way, to some extent, track the success of a product. These services evolve and improve all the time.

My advice is to ask the companies that provide these types of services to present what they can offer at regular intervals and keep up to date on new services that can improve your level of understanding. It is of course also possible to screen the online shop options available – either with manual searches or help from AI – a time-consuming task …

If you subscribe to new launch data services, you will get access to huge amounts of valuable data. These companies track new product launches globally. You can typically screen different markets and categories and compare what is trending. You will also be able to get information about the ingredients, nutrition, claims, pack design, price point, and much more. This information is very valuable to analyse, as it is, more or less, a transparent display of how the competition designs their products.

It can guide you in designing your product to have a point of difference. List all important factors in a table to get a good overview. Compare the ingredient lists of the closest competitors and compare them to your product. Compare the nutritional composition, the pack size, and the price point. What they claim on the front of the pack, and how do they describe it on the sides and back of the pack. This exercise will provide a great overview and can help guide the product design later in the process.

In the New Dawn project, you would screen out all the different types of RTD smoothies and get an overview of the key parameters. That data, along with consumer insights on what they seek to solve with a smoothie, can be used later in the process to help design the product. But let's look at another product category as an example, for the sake of diversity.

Snack Bar Example:

Imagine that you have a food production company that considers starting the production of snack bars. You already have a lot of the ingredients needed to produce a bar available in the company, as it sells oats, nuts, seeds, chocolate, dried fruits, et cetera. It is packed as single ingredients for food service. You have started an innovation project to clarify if snack bars would be a product type to bet on. One of the topics to clarify is how to position the product regarding the level of process applied, and the naturalness of the ingredients used. The company strategy and perhaps internal

TABLE 7.1

Variant & Company	Pack size	Retail price	Ingredients:	Nutrition Per 100g	Claims
Oat Chocolate Bar **Amazing Snacks**	4x 30g	£2.75 pc. 2.29/100g	Oats (38%), Chicory Root Fibre, Vegetable Oils (Rapeseed, Palm), Sunflower Seeds, Golden Syrup, Fat-Reduced Cocoa Powder (3.6%), Liquid Sugar, Chocolate (2.5%) (Cocoa Mass, Sugar, Cocoa Butter, Fat-Reduced Cocoa Powder, Emulsifier: SoyaLecithin), Humectant: Glycerine, Linseed, Palm Fat, Modified Starch, Soya Flour, Demerara Sugar, Starch, Natural Vanilla Flavouring, Sea Salt, Emulsifier: Soya Lecithin, Citrus Fibre, Natural Flavouring, Stabiliser: Xanthan Gum, Molasses	Energy: 1607 KJ 386 kcal Fat: 22g Carb: 33g **Sugars: 8.9g** Fibre: 17g **Protein: 6.3g** Salt: 0.23g	High in Fibre
Peanut Delight - Raw Fruit & Nut Bar **Natural Balance Foods**	4x 35g	£3.00 pc. 2.14/100g	Dates 53%, Peanuts 46%, Sea Salt, Natural Flavouring	Energy: 1776 KJ 425 kcal Fat: 21.5g Carb: 41.7g **Sugars: 38.8g** Fibre: 5.0g **Protein: 13.6g** Salt: 0.60g	No Added Sugar Source of Fibre

nutritional guidelines will provide a framework, but how will that look compared to what the consumers are already buying?

Table 7.1 includes two very different products to showcase how to build the overview. In reality, it would include 10 - 20 different products, highlighting similarities and differences. The document should also include a summary of those, and a recommendation on an obvious area where a new product could be designed with points of difference.

A summary and recommendation in this example could be something like the following:

We know from consumer research that users of snack bars find them convenient and tasty as an "in between meal," however, they have concerns regarding how healthy they actually are, due to the high amount of carbohydrates in many of the snack bars. The Peanut Delight Fruit and Nut Bar from Natural Balance Foods has a very simple composition of three ingredients. It has a high amount of carbohydrates, of which sugars at 38.8% due to the naturally occurring sugar in dried dates. It also has a high protein content of 13.6%.

At the other end of the spectrum, we have the Oat Chocolate Bar from Amazing Snacks. It has a very long ingredient list and could be perceived as highly processed and not very natural. The significantly lower carbohydrate content, with sugars at 8.9%, may appeal to consumers highly focused on sugar intake and health. However, the ingredient list and higher fat content could counteract their perception of its overall healthiness. A recommendation could be to place the snack bar in the middle of these two products. Aiming for a maximum of six ingredients, and carbohydrates, of which sugars, at 10%, from a dried fruit source, and a protein content of 8%. It will enable the company to design a more natural and not too unhealthy snack bar, with a delicious taste.

When conducting this type of product comparison, it is important to buy the products, taste them, and test them against each other! In the "Prototype chapter," you will be able to read more about the consensus mapping exercise, which is useful to map competitor products and their sensory attributes.

Online Store Check

Online shopping is a growing trend, and it is also possible to find a lot of useful information about competition that way. Finding a 3-year-old product launch, still available at online retailers, gives you a good indication of the product's performance. It is most likely not doing too badly, as food and beverage products that are not selling will most often be delisted. Here you will also be able to check the price in real-time and get the category overview for the specific retailer. It can also provide insights regarding further marketing communications aside from what is on the packaging. Some online shops provide additional information and pictures. If you, as a company, can't prioritize subscribing to launch data services, manual online checking is better than nothing. Especially if the competition is outside your home market, it provides "free" market insights.

Company Website Communication

In addition to the marketing communication that might be added by the online shops, visit competitors' websites and find the product you are analyzing, and review how they present and describe it. It all depends on the company's type and size, category, brand, market, and the extent to which its products are displayed. Non-existent websites, or poorly maintained pages, also reveal something about the company's priorities and strengths. If your close competition is a private label product, it is less likely to find anything.

Company websites often showcase their full range of products in the categories they have launched. This provides insight into the scale of their investment and the flavors they have prioritized for development. A range with more than three variants typically signals that the category is a high priority for them – likely because it represents good business potential. If the competitor is a large company, it is also possible that they have paid famous people to endorse their product, or maybe even be sponsors for a big event, like the Tour de France, or a famous music festival. All the information on a company's website reveals how they communicate their brand identity, who their target audience is, and for what occasions. This is especially valuable if their approach differs significantly from yours. Perhaps they are tapping into different emotional triggers. Are they reaching a different audience? Have they understood consumer needs better? These are key perspectives your company can reflect on. The website will often also hold information about the competitor's CSR policy – giving information on their stance regarding sustainability and responsibility.

Advertisement

Further information about communication is, of course, also to be found in the advertising campaigns for the competition. Campaigns are often used to boost product launches, leveraging a lot of different platforms. Advertising can come at many levels, but most companies will have a digital strategy and use ads on social media.

If the company and product are big enough for that type of investment, there can also be traditional TV ads and outdoor advertising, like billboards, transit advertising on buses, taxis, or in metro stations. Some companies also do street-level marketing with pop-up initiatives and handing out samples in crowded areas. Like a music festival or other big events. There can also be point-of-sales (POS) displays in the stores, which brings us to another classic approach. Check out the competition in the physical stores and observe what is on the shelves.

Store Monitoring

You will get a smaller picture compared to the subscribed type of launch data, but a much more tangible and vibrant experience. If you are a local market company, it is also an easy and cheap way to get a sense of competition. It's one of my personal favorite things to do! At home and especially when traveling. Visit a supermarket with the sole purpose of walking around the aisles and looking at the food and beverage products on display and the consumers buying them. Amazing! A big part of 2017 and 2018, I had the pleasure of backpacking around a big part of South-East Asia and Japan, with my wife and our two children. Mostly pure holiday and family time, but I also did some minor jobs here and there. One job I got offered was exactly a store monitoring job in all six countries we were planning to visit. I could hardly believe it and had to pinch myself when I heard about the offer. My response? "YES, please! I'm already doing it anyway." If your company exports to other countries, I recommend that core team members travel to those markets as well. This experience provides a much deeper understanding of the offerings and the consumers buying them. **This is an essential way to cultivate the empathy that is needed to fuel creativity later in the innovation process.**

Regardless of the location, pick out the most interesting retailers or foodservice distributors and visit them with your camera. Just use the one already built into your phone. It makes it less conspicuous. Some retailers are not too keen on people taking pictures in their shops. It is, however, my impression that it is not related to pictures of products, but retailers can, of course, be concerned about their competitors gaining insights into their store layout, product placements, promotions, or pricing strategies.

If you want to be on the safe side, ask the shop manager before snapping away. Also, be mindful not to take pictures of other customers or interfere with their purchase experience. But very important, observe them and make notes afterwards. In some cases, you can even ask the retailer if it is okay to ask their customers some questions and have a small open-ended interview ready.

When you visit a store, observe how the products are positioned. Is it the most intuitive part of the store to place this type of product? What products and companies have been placed in the best section of a shelf, cooler, or freezer? Top, middle, or bottom placement indicates how well it is selling, or how much leverage the brand has to get good placements. Middle placements, often referred to as "good eye-sight placements," are considered the best positions. Look at the category shelf of interest and take pictures of the most interesting products. Be mindful to include all the sides of the pack to ensure you have information on the pack design, the ingredient list, nutrition, claims, and so on. Also, take a picture of the price and the entire shelf to see the scale of the different brands and their position. The shelf pictures can be used to visualize how your brand and product ideas would look on the shelf. When you are at the store, take a look at some of the other aisles to see how different categories are presented and organized – it might be a great source of inspiration and learning. If you visit a store often, maybe where you do your shopping anyway, try to spot what products always seem to be on offer. It could indicate poor sales, but there may also be more positive reasons, such as ongoing promotions. You can determine if it's the first by checking the best-before date. If the products on offer are close to that date, it may indicate low stock rotation.

Consumer Shop-along

To provide rich insight to understand the dynamics of shopper behavior in a specific category, it is valuable research to ask consumers to do a shop-along with a camera to record their shopping journey. You can, as always, find specialized companies providing this type of research, including screening of relevant consumers. Consumers can use a handheld or wearable camera to document their experience and capture what they choose to buy and the reasoning behind their choices. They will have a tailored guide to conduct the shopping and recording. This method provides valuable insights into the following:

Product selection: What catches their attention first? What influences their final choice?

Competitive positioning: If you are already present. How does your brand stand out among competitors on the shelf, or fail to do so? What brand is the most noticeable?

Packaging and labeling impact: What elements help or prevent product recognition and appeal?

Store navigation and layout effects: How does the retail environment shape decision-making? Relevant info on where the best aisle placement is.

By analyzing these recordings, you can uncover patterns in consumer behavior and why the competition might have an edge, or not.

Social Media Listening

As pointed out earlier in the sub-chapter "Consumer Research," social media listening enables you to analyze consumer conversations, sentiments, and trends across social media platforms. This method can be used in competitor analysis as well. By tracking mentions, hashtags, and discussions, you can gain insights into how consumers perceive competitors and what drives the engagement. This method helps to uncover the following:

Brand perception: What are consumers saying about the competitors? Are the sentiments positive, negative, or neutral?

Consumer preferences: Which product's features, trends, or innovations are gaining traction?

Engagement patterns: What types of content drive the most interaction for competitors?

Pain points and unmet needs: What complaints or gaps in offerings emerge from consumer discussions?

Product Reviews and Articles

Analyzing product reviews and industry articles can also provide an understanding of consumer sentiment, product performance, and market positioning. By examining customer feedback on retailer webshops, brand websites, and review platforms, as well as expert opinions in articles, you can identify competitive strengths and weaknesses. This method helps to uncover the following:

Consumer satisfaction and dissatisfaction: What do customers praise or criticize about competitor products?

Key product differentiators: What features or benefits stand out in competitor offerings?

Recurring issues: Are there common complaints that indicate opportunities for improvement?

FOOD AND BEVERAGE TRENDS

With a good idea about the market situation and the strengths and weaknesses of the competition, it is valuable to look at more specific food and beverage trends. By looking at the products from the competitors and other markets ahead of the markets in scope, you will be able to analyse the products and the underlying behavioral trends driving them. This will provide valuable insight into where the pre-scope will be able to position itself and assess both the potential and risk of a launch. Food and beverage trends are, of course, a broader concept than just looking at the competition and categories in scope. In a wider context, any F&B company should monitor what is moving in all aspects of food and beverage consumption. What is served at the local street food market, at the hipster cafés, or the fine dining restaurants of relevant markets? What is the current social media F&B movie clips people obsess over? How is food styling presented, and what are the latest ways of combining ingredients and using clever cooking techniques?

F&B trends are evolving patterns in the way people consume, prepare, and think about food. These trends are influenced by various factors such as cultural shifts, health and wellness movements, technological advancements, environmental concerns, and changing consumer preferences. Food trends often reflect the broader social changes and values of a given time. F&B trends represent the dynamic nature of how we eat and interact with food and beverages. They are not only about what is on the plate, or in your glass, but also include the wider context of how food is sourced, prepared, and experienced. Food trends can span a spectrum, from ingredient preferences and cooking techniques to dining habits and even sustainability practices.

Here are two mega F&B trend examples. They are emerging from micro trends and becoming more established around 2010 And are still relevant in 2025, and most likely in many years to come.

Plant-based F&B Trend

The rise of plant-based diets reflects a desire for personal health, alongside the heavy influence of the media, governments, food companies, health experts, and environmental organizations promoting more sustainable food choices. This trend is evident in the growing availability of plant-based alternatives in supermarkets and restaurants, from meat substitutes to dairy-free products. Additionally, food companies and chefs are innovating with plant-based ingredients, creating new flavors and textures that appeal to a broader audience, even those who do not fully commit to a vegetarian or vegan lifestyle. This shift has also influenced fast food chains to offer plant-based burgers and meals, making sustainable eating more accessible to the mainstream.

The Fermented F&B Trend

Another example is the growing popularity of global cuisines, which reflects an increasing openness to diverse flavors and culinary traditions. A key driver of this trend is the rising appreciation for traditional food preservation techniques, particularly fermentation. Fermentation has gained significant attention for its health benefits, unique flavors, and deep cultural roots. This revival has sparked renewed interest in regional fermented foods, such as Korean kimchi, Japanese miso, and the global hit, kombucha. Kimchi, for example, has transitioned from a staple in Korean households to a globally recognized delicious superfood.

The widespread adoption of fermented foods aligns with the broader movement toward natural, minimally processed foods, as consumers seek out products with functional health benefits. The rise of fermentation workshops, artisanal fermented food brands, and increased representation in mainstream grocery stores further illustrate this shift. One of the most acknowledged authors about fermentation is Sandor Katz. He has popularized fermentation through books like "The Art of Fermentation"[11] from 2012, demonstrating how this ancient practice not only enriches flavors, but also strengthens local food cultures. The fermentation trend actually also ties into the sustainability movement, as it allows for food preservation without artificial preservatives, reducing food waste and encouraging seasonal eating. As we saw materialized in the business model by the Startup, "Reduced," from Copenhagen, adding value to food waste via fermentation.

THE F&B TREND CURVE

The impact of food and beverage trends is visible in restaurants, home kitchens, grocery store shelves, and in the ways food is portrayed on social media. A single viral dish, drink, or cooking method can quickly become a trend, shaping the way we experience food or drinks for a certain period. However, F&B trends are ever-changing, and what captures our attention today might evolve into something different tomorrow. By paying attention to F&B trends, you will be able to better anticipate consumer preferences and make informed decisions. While some trends may fade away, others become enduring staples, ultimately shaping the way we nourish ourselves and engage with the diverse and vibrant world of food and beverages.

To give you an overview of what is moving on in the specific category of your interest, it can be helpful to map products onto an F&B trend curve. This F&B trend curve is somewhat inspired by the bell curve from the "Diffusion of Innovations Theory" by Everett Rogers, and countless trend curves shown to me by companies and consultants throughout my career. This F&B Trend curve is different in the way that it helps to visualize how prevalent products are within a certain category and on a specific market. Additionally, it captures the primary emotional or functional need driving the trends. The placement of products should be based on market and competitor data, gathered by some of the methods described earlier. The drivers behind should be based on a diverse set of consumer research.

The structure of the overview is the following:

A. Choose a category or product type relevant to your project.
B. Pick the products and sort them into four sections:

1. **Emergent:** Products that are new to the market. Few consumers know about the product and what it is offering. Emergent products are often led by start-ups and smaller companies. It can, however, also be established brands with new niche products.
2. **Gaining Momentum:** Products that have been seen on the market for 12 - 24 months, with increasing availability and sales volumes. A good sign is also if the product starts to get media or influencer attention.
3. **Becoming Mainstream:** Products in the category that have been on the market for years and are to be found in a large selection of shops. These are brands and products that many consumers would recognize. Becoming

mainstream does not necessarily mean that the product is a top volume seller. The volume is also reflected in how often that type of product is normally used. Consumers will, for example, buy curry powder to use it maybe 2 - 3 times a month, but the stable muesli is consumed almost every day.

    4. **Everywhere:** Products by big brands are known to most consumers, even non-users.

  C. When screening a product category in a specific market, you need to identify what the driving trend has been behind the *becoming mainstream* and *everywhere* products, and how they correspond to the driving trends behind *gaining momentum* and the *emergent* products.

The underlying trends in a category or product type can remain consistent, with some being particularly strong and persistent. However, new trends occasionally emerge and begin to influence products. Analyze each product in terms of how it is communicated to the consumer. What the product is about, and what claims and pack design are used to support it? What behavioral trend is the most significant behind the product? To get a good visual overview, add the products onto an F&B Trend Curve, as illustrated in Figure 7.4.

As an example, we could take a closer look at the DK market on functional beverages 2025. Functional beverages have been a growing category for years, and in this case, it can give clarity to place some of the market products and see if there is a new trend within the category that is emerging or even gaining momentum. The functionality of the market leaders is mostly focused on the **fast energy** boost that

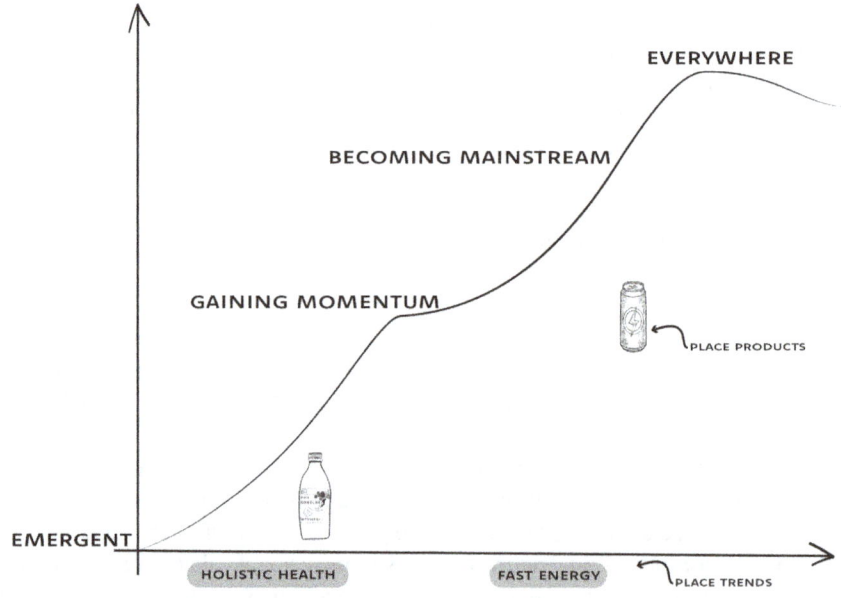

FIGURE 7.4   F&B Trend Curve.

their products provide, which directly connects to the needs of consumers seeking quick and effective ways to enhance their physical or mental performance. This behavioral mega trend is particularly prevalent among individuals with demanding lifestyles, such as professionals and athletes, but also among many young people in general, who are looking to boost energy levels to maintain productivity and focus. In this case, Red Bull and other major brands have been on the market for a long time, ensuring widespread availability. You can find them almost everywhere. The naturalness of the major players in the "Energy Drinks" market is generally low, as their formulations often rely on artificial ingredients. Many of these products are sweetened with artificial sweeteners. Additionally, they commonly contain synthetic flavorings and preservatives. To enhance their functionality, they usually also have a high caffeine content and are fortified with vitamins. The overall perception of these drinks remains tied to their artificial nature and functionality, and this has led to a growing segment of "natural" energy drinks that use and claim more natural ingredients. In the DK market, we can identify this trend by the significant growth of chilled RTD coffee products. These products have more naturalness associated with them, due to the natural caffeine content in coffee beans. They are often also not sweetened or sweetened with low amounts of sugar. This low natural sweetness appeals to health-conscious consumers.

These products are already mainstream.

So, what is emerging and gaining momentum in the DK market for functional beverages? An interesting emerging product type in Denmark is also a more natural, yet scientific, version of the classic energy functionality. A brand called Clutch® Nutrition[12] has products that talk about cognitive functionality and gut health. These are products that only use natural ingredients as botanical extracts and sugar. Isomaltulose, a naturally derived sugar, is particularly valued for its low glycemic index and slow energy release. Could it be gaining momentum? Only time will show.

Kombucha products are, on the other hand, gaining momentum in Denmark, with several players on the market. Kombucha is a functional beverage made from purely natural ingredients through an old artisan fermentation process. Kombucha is a fermented drink made from sweetened tea extract or herbs, sugar, and a symbiotic culture of bacteria and yeast, called a SCOBY. Potentially also with added fruit or vegetable juice. During fermentation, the yeast consumes the sugar, producing organic acids, live bacteria, and a slight natural carbonization. The result is a tangy and tasty beverage with potential health benefits of live bacteria and low sugar content.

Nordic Kombucha

I have interviewed Phine Katrine Kjaer Wiborg, co-founder of the brand "Nordic Kombucha,"[13] to understand how they started their business of developing, producing, and selling the old, but woke, beverage. Phine holds a degree in anthropology, specializing in food. She met the other two co-founders at DuPont Ingredients[14] in Aarhus during a project investigating the concept of "clean label"[15].

Rasmus Lybech Jensen and Flemming Vang Sparsø were both food engineers, and the three of them decided to start a company together. The aim was to start the production of a food or beverage product, tapping into the trends of 2018. Leveraging Phine's background, they used qualitative interviews and market research to identify fermentation, locally sourced ingredients, and organic farming as key trends. They also wanted to build on the formulation mindset of clean labeling, which brought them together in the first place. With that scope of trends and consumer understanding, they ideated on what type of product could hold the potential to fulfill those trends and consumer needs. They decided that kombucha was an interesting product to explore. At that time, kombucha was still fairly unknown to Danish consumers. In the Prototyping phase, they leveraged their engineering backgrounds and began exploring how to produce kombucha, with the constraints of using locally sourced organic ingredients. They didn't want to use black tea as traditionally described in recipes, and started experimenting with ingredients like the dried leaves of raspberry bushes – a very novel ingredient, indeed! They tested their prototypes with consumers, as well as potential customers, like speciality shops, cafés, and bars.

In 2019, they launched their first range of four products. Phine imagined that the main target audience would be young women, but six years later, she reveals that it is actually a slightly older part of the female population, who are the main consumers. Their main motivation to buy the slightly expensive kombuchas is centered around holistic health. Some of the key benefits the consumers, who actively provide feedback on their webshop, appreciated, are:

- Live bacteria
- Low sugar
- Natural ingredients
- Great taste
- No alcohol (Many of the consumers actually use the kombuchas as an alcohol-free wine/cocktail – another trend rapidly growing during recent years)

Some of the consumer feedback statements from their website are the following:

These are lovely, refreshing drinks with very different flavors. I mostly drink them in small portions to keep my stomach in balance, and that's exactly what I feel they do.

Fantastic taste and smells of health. Highest recommendations.

We love Nordic Kombucha. They are tasty, healthy, and a fantastic alternative to a good glass of wine.

I think the way a small startup like Nordic Kombucha worked with their innovation process is very much in line with the design thinking mindset. Despite limited resources, driven by deep empathy for the consumers, and a clear sense of emerging trends, they used creativity to innovate a product that truly solves unmet needs for certain segments of the population. Before the end of the interview, Phine also revealed that the new kombuchas in their innovation pipeline would include even more functional ingredients. They are building on a target audience of consumers, looking for natural products with even more **holistic health** functionalities.

PRODUCT EXPECTATIONS

The importance of product packaging cannot be understated, especially in the food and Beverage industry. Packaging creates the first impression and sets expectations as discussed in the consumer purchase journey.

PACKAGING EXPECTATION EXAMPLE

Figure 7.5 shows an analysis of the packaging design of a Sour Cherry Kombucha from Nordic Kombucha.

Freshness and Sensory Appeal

- The tiny water droplets on the bottle suggest the product is cold, fresh, and thirst-quenching. This visual cue immediately creates a craving for a refreshing, chilled drink. This image is, of course, only to cater to online shoppers, which is a very important distribution channel for the company.
- Glass bottles normally communicate premium quality, environmental consciousness,[16] and a connection to a handmade traditional fermentation process rather than mass production in plastic bottles.

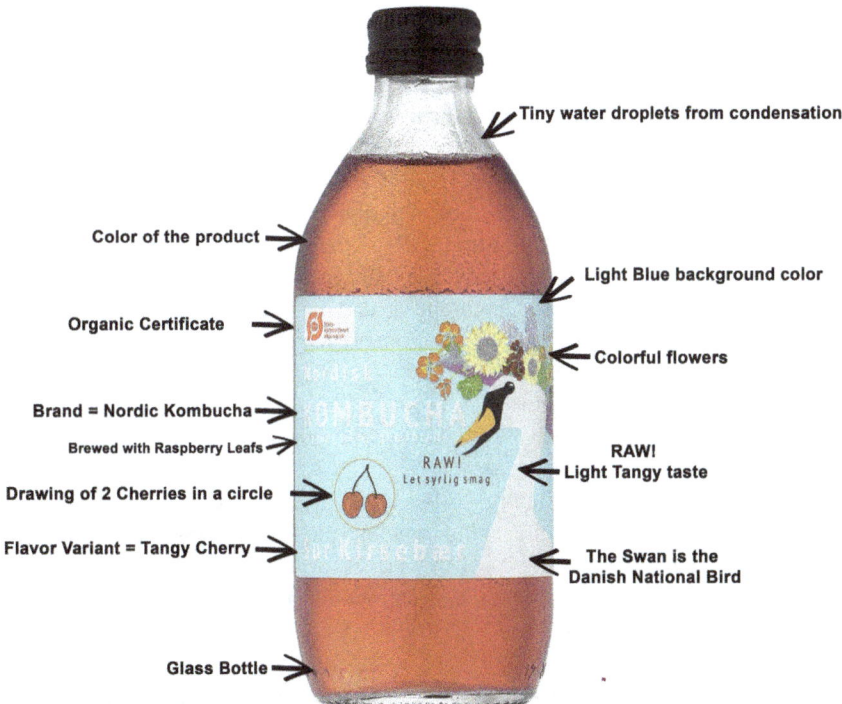

FIGURE 7.5 Nordic Kombucha. (Image courtesy of https://foodexplore.dk, used with permission.)

- The visible product color allows consumers to appreciate the natural, rich hue of Cherry Kombucha, reinforcing expectations of real fruit ingredients and an unprocessed, authentic beverage.

Natural and Organic Positioning

- Organic certification is a strong trust cue, indicating that the product aligns with holistic health values.
- Brewed with raspberry leaves, highlights a natural and novel botanical infusion, reinforcing the handcrafted and traditional process.
- RAW! Light and tangy taste. The word *RAW* emphasizes that the product is unpasteurized, with live bacteria inside.

Flavor and Taste Expectations

- A drawing of two cherries visually reinforces the flavor profile *Tangy Cherry*, making it easy for consumers to recognize the taste while emphasizing a crafted, illustrative aesthetic rather than a photographic representation.
- The use of the word *Tangy* sets the expectation of a slightly sour, refreshing flavor, which aligns with the natural acidity of kombucha.

Nordic and Cultural Identity

- The brand name *Nordic Kombucha* establishes a regional, premium identity with a focus on Nordic craftsmanship and quality.
- The illustration of the Danish national bird, the *Swan*, signals authenticity and Danish heritage.
- The light blue background color is actually the RGB: R169, G207, B213, which is a shade of pastel cyan. It falls in the soft, muted range of blue-green hues, often associated with tranquility and freshness.
- The colorful flowers cue naturalness and emphasize the use of botanical ingredients, and hence, a delicate, aromatic experience.

Overall Consumer Expectations

The design elements collectively set up expectations of what this Kombucha is:

- Premium and Handcrafted (glass bottle, organic certification)
- Refreshing and Thirst-quenching (condensation, visible product color)
- Natural and Healthy (raw, raspberry leaves, organic)
- Authentically Nordic (branding, swan illustration)
- Flavorful and Vibrant (cherry illustration, tangy description)

The design successfully balances aesthetics, functional cues, and storytelling, making it attractive to health-conscious, experience-seeking consumers, who value both taste and quality.

How to Use it

An analysis like this will provide you with several important insights to use further in the innovation process. Answering the question of what products are trending in the market, and why. This example was one of the product types gaining momentum, but the analysis should be done with the products that make sense in your project. A thorough pack design analysis provides a proposition and communication benchmark of competitors, helping to define a clear overview of areas where your innovation project could differentiate. It is an important part of the overall market and competitor analysis, covering both the dominant trends and those that are emerging and gaining traction. It also provides valuable inspiration on how to effectively communicate and ensure that product ideas and concepts resonate with a target audience, seeking to address both emotional and functional needs. More about this in the sub-chapter, *The graphical pack design.*

Visual Content Analysis

Lastly, it is of great value to collect new food and beverage trends spotted around the world. You can use hours following F&B influencers, get newsletters, look at online shops and restaurant menus, et cetera to stay up to date. Even better, get out there in society and shop, eat, and drink! And notice all the small changes and new ways of enjoying food and beverages. It is a great way to keep up to date with trends.

However, there are also faster ways to get the trend info if you need knowledge around a certain area for an upcoming project. There are companies doing the work for you and collecting all the trends in reports. This is great info to use later on in the ideation process for inspiration. Furthermore, it is possible to use companies that offer advanced AI solutions, which can analyze images and videos, identifying products, logos, and settings. This is especially valuable in product trend identification.

Market Players Overview

Any sensible company will also keep a close eye on who the other market players are within their category. Sales data can back up assumptions on how successful they actually are. Companies, in general, will always try to communicate great success, even if it is far from the reality of their revenue and profits. Get an overview of the market players and their success based on sales data, if possible. Also include facts like the following: Market dominance, the number of new launches in the last 2 - 3 years, and the price point. Perhaps you can even conduct a SWOT[17] analysis of the close competition.

Blue Ocean Strategy

Another way to approach competition is the well-known *blue ocean strategy.* In the book, *Blue Ocean Strategy,*[18] W. Chan Kim and Renée Mauborgne state the following:

The only way to beat the competition is to stop trying to beat the competition.

This is a bold and provocative statement! Their point is that attempting to create an improved version of the same products as the competition doesn't provide a true competitive edge. Instead, they argue for mapping out the industry and category standards to identify parameters where you can differentiate enough to significantly change the product offering in a way that attracts new consumers. They start by mapping out what the industry standards are. They call it a *strategy canvas*. With that overview in mind, they ask four questions: How can we:

1. Reduce beyond the industry standards.
2. Eliminate factors the industry takes for granted.
3. Create new factors never offered by the industry.
4. Raise factors above the industry standards.

Yellow Tail Wine Example:

Even though they never used the blue ocean strategy, the authors write about the Australian winemaker, Casella Family, and the development of their brand, Ye*llow Tail*[19] as an example. I think it is a great F&B example. They acknowledged that shopping for wine to some extent was too complicated for many consumers. And to be fair, all the different vineyards, grapes, vintages, production methods, and what to expect from the taste, can be very complex! They created a small selection of wines in 2001, branded *Yellow Tail*, with colorful kangaroos on the label. They raised three factors above the industry standard:

• Selection made easy (Clear and simple taste communication)
• Easy to drink (Mainstream taste profile)
• Fun and adventure (Colorful and fun packaging design)

All other factors, they eliminated or reduced. The price level was set a little above budget price but still significantly cheaper than premium wines. I recommend considering using the model in some cases. Maybe prepare a strategy canvas of the category standards for a workshop, and use the four-action framework in part of an ideation session.

The Past and Present

Examining the past and present of the F&B industry and the related consumer behavior offers valuable insights to inform and guide innovation projects. You might be able to anticipate even how certain trends will evolve and end up placing the new product innovation in the sweet spot of consumer liking. Successful innovation accomplished! But what about the outlook further on in the future? Let's look at the subtle art of crystal ball-looking … What might happen, and how can we prepare for it? While we may not aspire to become Nostradamus, the impulse to look ahead is embedded in the field of innovation. Identifying signals, imagining different scenarios, and preparing for uncertainty remain critical aspects of strategic thinking.

FUTURE FORESIGHT

Innovation would be so much easier if we could only predict the future! But as time-travel still is an innovation pending, we will have to live with the fact that nobody can look into the future and tell us what will happen … This has not prevented numerous people from trying to do so throughout time. The motivation to predict the future can be many: everything from economic gain to risk management can be motivations to attempt. Think about all the predictions around climate change. The first predictions around climate change were actually formulated in the early 19th century, and to this day, the predictions have changed decade by decade. The climate change predictions also show how hard it is for mankind to believe predictions. Also, in literature and movies, prediction of the future is such a big topic that it has its own genre: Science fiction. Personally, I am still waiting for the *hoverboard* to become commercially available. The following was predicted in the movie:

Back to the Future, Part II

It was released in 1989, and the filmmakers predicted that flying skateboards, or *hoverboards*, would be part of everyday life by the year 2015. This is not a prediction that came through. If you take a look at the iconic science fiction movie ***Blade Runner***, released in 1982, directed by Ridley Scott, some of the predictions made in 1982 did actually come true, even though the world in 2019, where the future scenario played out, didn't look anything like the actual world of 2019. Still, some of the predictions did come true. Digital billboards, voice-controlled technology, and in 2025, the point in time where these sentences are written, the reality of generative artificial intelligence is influencing our lives to an unseen extent!

Therefore, the question is whether it makes sense to try and predict the future in the food and beverage industry. The answer is very clear: YES, we should! The next logical question would be: Why? It is not possible to predict 100% what the consumers will talk about and find important five years into the future! That is clear! But to have some qualified guesses on how society and the consumers will evolve in the future is a timely action to investigate and subsequently assess future potential opportunities for your food and beverage company.

It can help the food and beverage company to do the following:

- Shape the company strategy.
- Stay competitive by proactively working on more long-term projects. These should be projects where it makes sense to look at more radical innovation within the food and beverage industry.

The next question one should ask is whether one should use generative AI to help predict the future? The answer is a definite YES, we should!

Until recently, this area only focused on traditional market research on a macro and micro trend level, with *human intelligence* as the predictor. With the growing capabilities of generative AI, it will be exciting to see how the combination of human intelligence and artificial intelligence will shape and develop the future of future foresight.

There are several indications that it will make the forecasts more accurate, as the AI tools can digest and make sense of huge amounts of data. And in a very short time, compared with human intelligence!

The amount and variety of data that can be incorporated into the research is astonishing! It can include the following:

- Usage and attitudes research reports.
- Segmentation profiles, including drivers of liking and daily life mapping.
- Other types of relevant research reports.
- Conversations on social media, blogs, and other online forums. (Especially, chef and bartender forums can contain interesting conversations on new trends within food and beverages.)
- Articles published in news media and other media.
- Scientific papers of relevance to the topic.
- Ratings and reviews on products and services.
- Search data from relevant platforms as Google, YouTube, TikTok, et cetera.

Research companies will provide these solutions in many qualities, and as a food and beverage company, it is all about finding a good match with a research company that has the capabilities – at a reasonable price.

Let's look at further initiatives to consider in future foresight work.

Ethnographic Research

Ethnographic research can create powerful insights into future foresight because it provides a deep understanding of consumer behaviors, preferences, and cultural influences within real-life contexts. By observing consumers in their natural settings, you can uncover insights about eating habits, decision-making processes, social influences, and unmet needs that surveys or traditional market research might over-look. "Don't ask consumers what they do, watch what they do," is a core principle of ethnographic research. Often, consumers aren't fully aware of their own habits, or they may respond based on what they think is socially desirable, as described in the passage about "The Say-Do Gap" in the consumer research sub-chapter. In the book, *Hidden in plain sight*: How to create extraordinary products for tomorrow's customers,[20] the author, Jan Chipchase, writes in his introduction:

A large part of my job is to spot and decode the little things that most people take for granted.

Hence the title, he alludes to all the small things people do that they do not neces-sarily think much about, or that seem to have a lot of impact on their lives. But it is the small maneuvers that people do, or change to solve small problems, or to handle new situations, that potentially could be identified as an unmet need. An unmet need that could have future potential to be solved with an innovation. Ethnographic research typically involves observing and engaging with consumers in places where food and beverages are bought, prepared, or consumed. Researchers will take detailed notes, record interviews, and sometimes filming interactions to capture context, emotions, and nuances. They may ask open-ended questions, allowing consumers to share stories and insights in their own words. This is a discipline that requires time and resources and should normally be allocated to projects of strategic importance and purely future-foresight explorations. It also goes without saying that not all companies will have the size and resources to enable this type of understanding work.

Expert Interviews

If you want to gain cutting-edge insights in a strategically important area for your company, it makes sense to talk with experts in that field. If you want to know more about the newest products, diets, and types of training that offer specific health benefits, talk with experts within the field of diet and exercise. In that case, it would be informative to investigate how a sport like professional cycling works with nutrition and restitution. An approach could be to interview nutritional experts working in professional cycling. The athletes in biking are working with an extreme consumption of calories and put their bodies through intense efforts. In a race, they will push their physical limits, requiring intense endurance and strength. The next day, yet another stage in the race, which will require very fast recovery efforts to be ready again. How do they work with that? And what ingredients, measures, or techniques do they use? Experts can also come directly from academia and bring the latest research data and knowledge within a range of areas. It can be a professor in psychology with expertise in how humans think, or an anthropologist with knowledge about different cultures and human behavior.

Professionals working in a relevant sector in society can also help inform and translate some of the issues people will have more of in the years to come. It could be a nurse, a taxi driver, a hairdresser, a shop assistant, teachers, and others who don't see themselves as experts, but many times are real experts in human behavior. Next-generation ingredients and technology experts are, of course, also relevant when looking into the future of food and beverages. They can tell you about emerging technologies and ingredients that will enable new ways to produce products. They may be new taste experiences, new health benefits, new shelf life, new cooking functionality, and whatever they can come up with. These experts might already work within your company, or with a company which you collaborate with. It can include product developers, application specialists, process designers, scientists, and so on. If they are working at a company that wants to stay competitive, they will most likely already be working on several explorations – maybe novel and wild product ideas that might or might not have potential as the "next thing!" All serious companies should leave room for this type of sandbox exploration work!

First Movers

The first mover consumers, as earlier described, can be a valuable asset in future foresight as well. It could be the very curious consumer who has always tried the newest thing. It can be within a certain area as food and beverages, but also other areas are interesting. It could be consumers who are the first to try new digital solutions, new kitchen tools, gadgets, or transportation. (As when the first hoverboard will be launched!) You can benefit from conversations with these *first movers* but also have them record their lives in a video diary. It can also be more professional *first movers*,e.g., the **Influencers** that are often first mover consumers, but with the professional add-on that they had already posted about their experiences. Some have more commercial profiles than others. Skilled **chefs** and **bartenders** are obviously on top of the new trends within the culinary scene, and they can have very hands-on experience with the latest changes in taste preferences. They know about the new techniques used to create creative taste experiences.

Product developers and product **designers** are also often on top of new movements within their field. It can be any field that works with designing for consumers' unmet needs. Creative people like **artists** and **authors** can, in some cases, also bring new perspectives on how people's lives will evolve.

The conversations with the more professional first movers can be solo interviews to inform you about their behavior, observation, and predictions.

FIRST MOVER INCUBATOR

Imagine you ask a diverse group of first movers and experts to be part of your company's future foresight panel. Bring them together once a year and have them explore topics in a future sandbox environment. They would engage in an exploratory collaborative challenge, not bound by strict constraints or predefined outcomes. The starting point could be a video from a first mover, capturing an interesting topic or a science fiction movie. Ask questions like the following: What if this behavior becomes more prevalent in society? Ask them to brainstorm and anticipate challenges and potential solutions, focusing on areas like digital advancements, culinary experiences, or consumer lifestyle shifts. It could be a "roundtable setup" where the participants reflect in turns and identify collective insights, pinpoint surprises, and reinforce what might become strong trends. Capture the conversations and foresights to be included later in a long-term innovation process. Honestly, this type of setup is not used to a large extent in the food and beverage industry. But maybe it should! It sounds very ambitious, but it can be a way to be proactive as a company and differentiate your level of future foresight from the competition. The setup can be configured in many fidelities, depending on your company's size and resources.

"The best way to predict the future is to create it!"

This is a quote widely attributed to Peter Drucker,[21] but regardless of who said it first, I believe it to be very true and certainly a mindset to be embraced in food and beverage innovation, as well. Always leave room for future foresight, sandbox explorations, and the occasional bold launch! It enhances the creative energy within a company and keeps it interesting for all. The employees, the collaboration partners, the customers, and of course, most importantly – the consumers.

OUTCOME OF THE UNDERSTAND PHASE

This marks the end of the Understand phase in this book, but in reality, it is a phase you will likely revisit throughout a project. New knowledge and consumer empathy will enable the project participants to ask new questions and challenge assumptions. Going back to understand more is a core mindset of design thinking.

Back through the waves to catch a new wave, wiser and ever curious.

It is NOT recommended to initiate all areas of research and methods, as doing so would be too time-consuming and expensive overall. And I am not suggesting that all companies need to operate like research companies themselves, either. Gathering the right knowledge requires expertise, and it may be more effective to collaborate with one of the many supporting companies in the F&B industry. The finalizing of all the selected types of research and writing them into summaries and reports is the last task in the Understand phase. Create an overview of all the learnings and prepare for the next phase, the Define phase. The collection of documents should include the following:

- Pre-scope document.
- Target audience (segmentation).
- Reports from targeted consumer research. The outcome can be a report on consumer "Usage and Attitudes" (U&A), often provided by market research companies to document the exploration of consumer behavior, preferences, attitudes, motivations, and perceptions regarding products or services.
- Target audience occasion journey map.
- Job to be done analysis.
- Market and competitor analysis, including F&B trends.
- A summary that outlines the objectives, key findings, and project implications.

Present the results to the core team and have a productive discussion about the implications and opportunities to build on as the project moves forward into the next phase: **The define phase**.

NOTES

1 https://youtu.be/WGuVEVvB52c?si=givkJNlZhAf_DVcu
2 *The Creative Act: A Way of Being*, London, Rick Rubin, Penguin Press, 2023.
3 *Effective Brand Building: Unlock Growth with Strategy, Insights and Measurement*, Andrew Geoghegan, Kogan Page, London, 2025, Page 85.
4 *The Consumer Insights Handbook: Unlocking Audience Research Methods*, Danielle Sarver Coombs, Rowman & Littlefield Publishers, Maryland, 2021.
5 *Consumer Behavior* (Fifth edition), Zubin Sethna, Sage, New York, 2023, Page 145–146.
6 www.perspective-lab.com
7 *Competing Against Luck: The Story of Innovation and Customer Choice*, C. M. Christensen, K. Dillon, T. Hall & D. S. Duncan, Harper Business, New York, 2016.
8 *Identity Designed: The Definitive Guide to Visual Branding*, David Airey, Rockport Publishers, Gloucester, MAs, 2019.
9 This framework draws inspiration from established marketing and consumer behavior models, including E. St. Elmo Lewis's AIDA model, McKinsey's Consumer Decision Journey, and principles of customer lifecycle marketing.
10 www.bostonteapartyship.com/tea-blog/who-invented-the-teabag

11 *The Art of Fermentation*, Sandor Ellix Katz, Chelsea Green Publishing, Chelsea, Vermont, 2012.

12 https://clutchnutrition.com/

13 https://foodexplore.dk/

14 Today a part of IFF - www.iff.com/

15 The term *clean label* is used in product development, basically aiming to develop and produce products with an easy-to-understand ingredient list. Free from artificial additives, preservatives, or ingredients with complex chemical names. E-numbers are also generally viewed as undesirable, due to consumer perceptions. Even though many E-number additives are safe and naturally derived, companies often remove them or rephrase them with more familiar names to align with consumer expectations for simplicity and naturalness.

16 Sustainable glass bottle handling in Denmark.
    In Denmark, you pay a small deposit on bottles and cans, which is refunded when you return them. Glass bottles are reused many times through this system, making them more sustainable than in countries where bottles are only recycled after a single use.

17 The widely used strategic planning tool for assessing Strengths, Weaknesses, Opportunities, and Threats (SWOT - Commonly attributed to Albert S. Humphrey).

18 *Blue Ocean Strategy: How to Create Uncontested Market Space and Make the Competition Irrelevant* (Expanded Edition), W. Chan Kim & Renée Mauborgne, Harvard Business Review Press, Brighton, MA, 2015.

19 www.yellowtailwine.com/

20 *Hidden in Plain Sight: How to Create Extraordinary Products for Tomorrow's Customers*, Jan Chipchase & Simon Steinhardt, Harper Business, New York, 2013.

21 Consultant and author Peter Drucker wrote several influential books, such as The Practice of Management (1954) and Management: Tasks, Responsibilities, Practices (1973).

8 Define (Frame It)

CONSTELLATIONS OF POTENTIAL DEMAND (CPD)

In the Define phase, it is all about converging. We are the surfer swimming back through the waves with all the learnings from the previous waves, ready to condense the knowledge into the next wave.

The aim is to narrow down all the learnings from the Understand phase and translate them into different constellations of potential demand.

Use the desirability lens and ask essential questions.

- Who could be the target audience of your project?
- Is the project seeded in a food or behavioral trend? Describe and validate the trend.
- Why would the project resonate with the emotional and functional needs of the target audience?
- What is the occasion? When and where would the consumer use the product, and what motivates its use in that specific context?

The different constellations of potential demands, based on the pool of knowledge gathered, can be assembled in many ways.

Cluster the knowledge into five areas:

1. Who
It may be different constellations of consumer segments, a variety of age groups, family status, lifestyles, life stages, mindsets, attitudes, and behavior.

2. Trends
These consumers are influenced by certain trends. Macro and micro trends, specifically, are relevant in the food and beverage industry.

3. Why
Functional and emotional needs and tensions experienced by the consumers in their daily lives as well as the motivation to act on them.

DOI: 10.1201/9781003619352-10

This chapter was refined for grammar and fluency using ChatGPT-5.0.

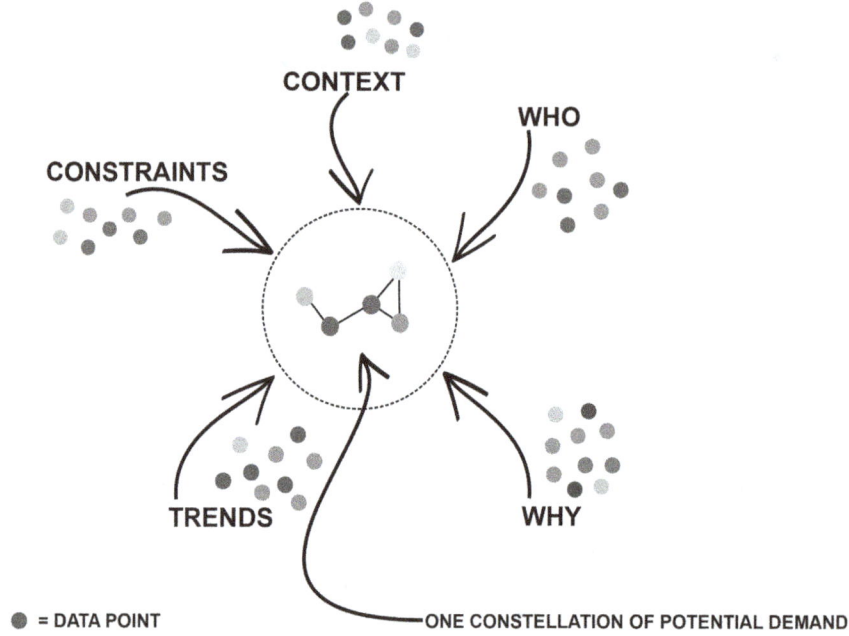

CONTEXT

WHO

CONSTRAINTS

TRENDS

WHY

● = DATA POINT   ONE CONSTELLATION OF POTENTIAL DEMAND

FIGURE 8.1   Constellations of potential demand.

4. Context
The people they are with, if any, and the locations where these moments occur.

5. Constraints
It is also possible to add some inside-out constraints or wishes to the project scope, Like the feasibility lens. It could be a certain product category that is important to the company, or a valued brand that is prioritized. Also, look for overlapping, complementary data points and logic connections for each area.

Use curiosity and knowledge to join the dots. Ask plenty of questions and have in-depth discussions in the core team. Make sense of all the input, and cluster it into constellations of potential demand that are meaningful and coherent. See the illustration, Figure 8.1 – there can be several constellations within the same pre-scope of a project. Gather as many as possible. The constellations will indicate different angles of desirability.

OPPORTUNITY SPACES

Each constellation of demand opens up a potential space of opportunity, which can be framed as *Opportunity Spaces*. Assess all the constellations of demand and build on those that gather the highest consensus.

The work to identify the potential constellations of demand and the translation into opportunity spaces should be done by the core team. It could be facilitated by the project leader or an external facilitator. In preparing for the final project scope, aim to create 2 – 4 opportunity spaces to maintain a balance between focus and exploration. These spaces are designed to define the final project scope and serve as a foundation for ideation. However, not all opportunity spaces may ultimately be utilized, as the process should be flexible enough to ensure alignment with the project's objectives. In some cases, only one might be relevant.

The final selection will depend on the pre-scope and how well the identified spaces align with the project's objectives.

Essential Questions for Opportunity Space Framing:

1. For who are we solving?
2. What is influencing the persona?
3. What tensions should be solved?
4. When are the tensions most prevalent and relevant?
5. What are the jobs to be done by the product or service?
6. What is hired today? Why it is hired and what barriers it might have?
7. What is the essence of the problem to be solved, in an "insight sentence"?
8. How can the insights be rephrased as a "how might we" (HMW) question?
9. What are the potential project constraints?
10. What is the title of the opportunity space?

Answering these questions and formulating the potential demands into opportunity spaces will bring a lot of clarity to the team. The opportunity spaces form the foundation for building empathy among all participants as we move into the creative phases of innovation. They are a cornerstone of the design thinking mindset and process. **N.B.:** If the constellations of demand don't translate into opportunity spaces that evoke excitement and build reasonable volume expectations within the core team, it could be that you need to go back in the process to understand more, before finalizing the Define phase.

Let's assume it is not the case, or the trip back of catching new waves of knowledge that has been completed – and exemplify the details of how an opportunity space is created.

What could it look like for The Golden Juice Company and the innovation project – "New Dawn"?

Remember the questions from their pre-scope?

- What are the behaviors and challenges connected to a busy lifestyle, that is still aspiring to be healthy?
- What potential target audiences would be in scope?
- How do their mornings look like, regarding the "on the go" moments?
- What are the usage and attitudes towards smoothies today?
- We have seen more and more veggies appearing in new products from the competition. Should we also go that way?

The following three examples could be potential constellations of demand, discovered in the Understand phase of the project. The first opportunity space will serve as an example to explain the topics to include in an opportunity space. The last two will be additional examples without the explanations.

OPPORTUNITY SPACE 01: BUSY MORNING FUEL

Who are we Solving for?

This section introduces the persona, ensuring the team has a clear understanding of the target consumer. The persona is normally not a real person, but a fictional character built on the consumer segmentation relevant to the pre-scope, and insights gathered from consumer research conducted with the target audience. To make the different spaces relatable and to allow the workshop participants to empathize with them, a detailed narrative about the persona should be crafted. Include demographic details, lifestyle, and key characteristics. Emphasize their goals, motivations, and challenges – specifically those relevant to potential occasions. This will set the foundation for the ideation process. Additionally, enhance relatability by including a picture of the persona and giving them a catchy, memorable name. The photo can be generated using AI, sourced through an image search, or even drawn by a good sketch artist. Figure 8.2 illustrates the capabilities of AI as a fast, effective, and currently inexpensive way to capture the essence of a persona and I have chosen this method to visually bring the persona, *Career Carrie*, to life.[1] This shows its abilities in mid-2025, and the assumption is that it will continue to improve in the future. All the information collected about a persona from the research can be used to prompt the software of your choice on how to visualize the persona.

Persona: Career Carrie

Carrie is 38, a resilient and ambitious professional, balancing her demanding career and family life. She lives on the outskirts of London in a spacious home with her partner, Paul (41), and their two children, Emily (10) and Jack (7). Carrie works as a lawyer in London. A role that requires precision, fairly long hours at times, and constant focus. Her commute takes her from home to the children's school, and then to the bustling city center. While her profession is time-consuming, she is deeply committed to her family's well-being and aims to provide them with the best life possible. Carrie's lifestyle reflects her dual priorities of professional success and family harmony. Her home is well organized yet lively, with a mix of modern comforts and a warm, inviting atmosphere. She thrives on structure and relies on careful planning to balance her responsibilities. Her mornings are often hectic, starting with getting the kids ready for school and packing their lunches, before heading into the city for work. Once at the office, she juggles meetings, case preparation, and client calls, often skipping meals due to her packed schedule. Evenings are dedicated to her family, helping with homework, and winding down with a brief moment of relaxation before preparing for the next day. Carrie values practicality and is pragmatic about the challenges of her busy life. While she strives for balance, she acknowledges that compromises are sometimes necessary. She seeks convenient solutions that help her

FIGURE 8.2   AI-Image generated with OpenAI, ChatGPT-5 (August 2025).

maintain her energy and health without taking away too much time from her work or family. Carrie believes in the importance of healthy living, but is realistic about her limitations. She looks for high-quality time-saving products that align with her goals of efficiency, nourishment, and simplicity.

What is Influencing the Persona?

Here, you provide context for the persona's decisions and behaviors. This includes both external factors, such as cultural, social, or economic influences, and internal drivers, such as values, habits, or emotional states. Broader trends or environmental factors, like increasing health consciousness or a fast-paced lifestyle, can also be included to give depth to the understanding of the persona.

Trend:
The trend of shifting eating patterns:
Busy lifestyles, like Carrie's, have prompted a shift in eating patterns and behaviors. As a lawyer managing a demanding career and caring for her two young children, Carrie often prefers snacking and "on the go" eating habits over traditional sit-down

meals. This trend has fueled the demand for portable, handheld, and single-serve food options that can be consumed quickly and conveniently, while on the move. Perfect for someone like Carrie, who values efficiency and practicality in her daily routine.

Food habits:
Overall, Carrie's food habits reflect the challenges of balancing a demanding career and family life, often leading to missed meals and relying on convenience options when time is tight. She tries her best to make healthy choices within the constraints of her schedule and circumstances, recognizing the importance of nourishing her body to support her busy lifestyle.

Tensions and the Motivations to Resolve them?

What are the tensions to solve, and how motivated is the individual to resolve them? This question identifies the conflicts or unmet needs that create opportunities for innovation. Focus on core tensions – when a persona wants to achieve something but faces barriers. Highlight both functional tensions, like time constraints versus health, and emotional tensions, such as guilt versus convenience. And the crucial question is whether the person feels motivated to act on it.

Why:
Functional Tension:
Carrie experiences a tension between excelling in her demanding career and spending meaningful time with her family. This dynamic leaves little time for meal planning, preparation, or even eating – often leading to a drop in energy and focus by mid-morning at work. It is a significant tension for her, and she is motivated to make a change.

Emotional Tension:
Carrie deeply desires balance in her life. She wants to provide nutritious meals for herself and her family, while also succeeding in her career and being present for their children. The emotional tension lies in the ongoing struggle to find this balance. A challenge that weighs on her, both practically and emotionally.

When are the Tensions most Prevalent and Relevant?

Identify the key moments or contexts when the persona experiences the tensions most acutely. These could be specific times of the day, like hectic mornings for busy professionals, or broader life contexts, such as meal planning or commuting. To recognize these moments helps to narrow the focus of potential solutions. Based on the consumer occasion journey of Carrie, we know certain details about her context.

Context:
Morning:
Due to her hectic schedule and morning rush to get the children ready and drop them off at school, Carrie often finds herself skipping breakfast before starting her commute.

Mid-morning:
As Carrie settles into her work at the office, she begins to feel the effects of skipping breakfast, as her energy levels dip. She reaches for whatever quick snacks she has on hand, if any.

Lunch:
Carrie sometimes has meetings at work during lunch time, which makes the energy situation even worse!

Jobs To Be Done (JTBD)
Remember the jobs-to-be-done-theory described earlier? Frame the JTBDs in functional, emotional, and social needs. Clarify what the solution needs to accomplish to address the persona's needs effectively. This includes functional jobs, such as providing quick energy.

Emotional jobs, such as supporting guilt-free choices.

Social jobs, like enabling the persona to share or recommend the solution confidently.

Lastly, how important *taste* is for the persona in the context of needs and occasions.

JTBD's for Career Carrie:

Functional:

- Nothing to prepare.
- Convenient to bring to work.
- Provide energy.
- Fast and not messy to consume.

Emotional:

- Feeling healthy.
- Feeling sharp.

Social:

- Like to share the product experience with others.

Tasty:
Taste is very important for Carrie due to these factors.

- If the taste is unbalanced, it undermines the functional need for quick and pleasant consumption. When a product tastes bad, consuming it can become a struggle. (At least if the product is over bite or shot size.)
- If the product fits the taste expectations, it will also support the emotional needs. In this case, the taste should signal a healthy product. (For most

TABLE 8.1

Hired product	Why	Barrier
Energy drink	Easy to transport Provides energy	Doesn't satisfy hunger Artificial taste Feels unhealthy
Fruit Smoothie	Fast and convenient Sweet and tasty	It has a high carbohydrate content that affects blood sugar balance and is not very filling
Fruit	Healhty Tasty	It can be messy and noisy to eat Can get bruised in the bag
Musli bar	Tasty Easy to transport Don't need cooling	It has a high carbohydrate content that affects blood sugar balance

consumers, it would mean that it shouldn't be too sweet, but still with a good and balanced taste. That goes for Carrie as well.)

If the product is tasty and fulfills her needs, it is also an experience that Carrie would like to share in her social networks.

What is Hired Today?

The "what is hired today" exercise can furthermore help to get an overview of how the tension is solved today. Explore the persona's current behaviors by examining the products or services they currently use to address the job. Ask what products *Career Carrie* would hire today to do the job? These answers should be available, looking at the consumer research conducted and the competitor analysis. Some of the products they hire today may be very good solutions. Some products might be very similar to what your company can produce.

Additionally, try to identify why these solutions are chosen, whether it is for convenience, taste, availability, et cetera. Highlight any barriers, such as being unhealthy, messy, expensive, or more and, as previously mentioned, look at the barriers and think of them as opportunities to solve for the consumer. This analysis can uncover gaps or opportunities for improvement. In Carrie's case, it could be as displayed in Table 8.1.

The Essence of the Problem to Solve

Try to condense all the insights into a single, clear, and engaging sentence that highlights the core tension and problem to solve. Such an insightful sentence should be easily understood. Here are four questions to circle in on what an insight is about:

1. *When?* = the time of day or specific occasion
2. *Persona wants?* = desirable outcome or need
3. *But?* = specific tension or barrier
4. *leading to?* = consequence of the unmet need

In the case of Career Carrie, the essence of the problem to solve could be formulated as follows:

Carrie wants to start her day feeling energized and focused, but her demanding morning routine often leaves her skipping breakfast, leading to low energy later on in the day.

This example is built by starting with the desire to be energized and focused at work. It is framing the morning situation. The *but* is used to describe the tension between the need and the time constraints of her life. The sentence ends with the consequence of the unmet need, namely *low energy*.

N.B. Don't get too hung up on using these specific words and refrain from getting into nitty-gritty discussions about how to formulate these. As long as the sentence captures the essence of the problem, it is fine. The advice is based on countless experiences of wasting time in discussing small iterations of formulations. That is not productive, and a good project leader or facilitator needs to handle this in a sensible way.

How Might We …? (HMW)

To further build on the insight sentence and make it actionable for ideation, use the "How Might We" (HMW) question framing technique.[2] This technique is used in the design thinking methodology to open up the problem space and encourage innovative thinking. It can help to unlock the opportunity space and transform the identified challenges into problems to be solved. By starting with "How might we," the question sets a positive explorative tone, promoting an inclusive and collaborative approach to brainstorming and problem-solving. Don't make the HMW too broad. Try to address the need in the opportunity space and some of the jobs to be done. Also include the target audience. It could sound something like the following:

How might we create a convenient, great-tasting solution that provides sustained energy and fits seamlessly into Carrie's hectic mornings?

This HMW is specific to Carrie's Persona and acknowledges her busy routine and prioritizes convenience. It emphasizes potential benefits like *great tasting* and sustained *energy*. Great taste is more or less mandatory. Convenient and sustained energy directly addresses her pain points. It opens the door to a wide range of ideas without being too prescriptive. If there are any constraints, they can be captured in the HMW, but it is recommended to keep it as open as possible.

What are the Potential Project Constraints?

Depending on the company strategy and weight of incremental versus radical innovation aspirations, this is where to state any limitations or wishes that might impact the development of the solutions. These could include limitations like production capabilities, "time to market" requirements, or budget constraints. It can also be wishes, like a strategy focusing on a certain category, a core market, an occasion, or sustainability goals. In the innovation project, *New Dawn*, the constraints are the following:

- **Category:** Potentially some form of smoothies in a bottle.
- **Occasion:** Starting from breakfast until lunch as the main focus.
- **Markets:** United Kingdom, Germany, and The Netherlands.

What is the Title of the Opportunity Space?

Conclude with a concise and memorable title that reflects the persona's core need or aspiration. For example, a title like **"Busy morning fuel"** encapsulates the opportunity space in a way that is easy to communicate and enables fast focus from the participants.

Other Examples

Other examples of constellations of demand can be many, and hence the following opportunity spaces would have different perspectives and possible solutions. One way to explore CPDs is by tracking the same persona across different points in their occasion journey. For example, if we consider Career Carrie, we might analyze her afternoon routine, including her mid-afternoon coffee break at work and her commute home. Each moment presents different needs and touch points that could be leveraged for innovation. Another approach is to examine a similar persona at the same stage in the journey, but within a different market or cultural context. For instance, Career Carrie in Berlin might have different expectations and constraints when grabbing a quick breakfast compared to Career Carrie in London. Local infrastructure, product availability, and cultural attitudes toward convenience and health would influence her choices, thereby shaping different CPDs. Or, a CPD can be explored by looking at a different persona with different values and needs, at the same point in the morning routine. Even though the occasion of starting the day is the same, the way it unfolds reflects personal priorities, responsibilities, and lifestyle choices. These different perspectives can reveal unique opportunity spaces, leading to different solutions in some cases, while in others, the same solution may work across audiences and their needs. Ultimately, the breadth of CPD exploration depends on how broad or how focused the pre-scope is. A narrow scope might target a specific product development within a single scenario, while a broader one could explore macro shifts and emerging behavioral patterns across multiple regions and touch points.

Atmosphere and Overview

Fill the presentations of the opportunity spaces with more detailed information, but limit the detail to what is actually relevant in selected spaces. Create presentations that are easy to understand, engaging and fun. It is a good idea to create a single template to structure all opportunity spaces, allowing for a quick overview and clear presentation of the key points. It can be a digital whiteboard version, but it could also be very powerful to create a big poster of each opportunity space and hang them on the wall. They will be used extensively in the final project, scoping exercise, and in the potential gate meeting, where it is decided which opportunity spaces to take forward in the innovation project. Later in the project, the selected opportunity spaces will again play a significant role in creating the needed empathy to enable creativity, leading to the end goal of a successful innovation.

ADDITIONAL OPPORTUNITY SPACE EXAMPLES

As mentioned, two additional examples of different constellations of demand and extracted opportunities will follow now. To broaden the example of the New Dawn project, we will explore other potential target audiences, different contexts, and the influence of other trends and motivations. The next opportunity space example features a different persona type, this time living in Germany. Even though the persona is created as a male living in Berlin, it could be based on consumer research across all three core markets for The Golden Juice Company. Target audiences that can be identified across markets are key to ensuring the viability of multiple markets or global launches.

OPPORTUNITY SPACE 02: VEGGIES MADE DELICIOUS

Persona: Family Franz

Franz is 45, a thoughtful and intentional family man balancing his role as a university professor, husband, and father. He lives with his wife Anna (43) and their three children, Lena (17), Clara (15), and Max (12). They live in an apartment in the vibrant Berlin neighborhood of Kreuzberg. Their home features energy-efficient systems, sustainable materials, and a small rooftop garden where Franz is trying to grow herbs and vegetables. Franz leads a balanced life, blending family activities with his passion for sustainability. Weekends often include family time, cycling through Berlin's green spaces, and tending to the garden. Franz values a healthy lifestyle for himself and his family. He is also concerned about the planet's future for his children and is trying to make sustainable choices. His evenings often revolve around cooking plant-based meals using fresh, organic ingredients from local farmers, and even sometimes the rooftop garden. Despite his commitment to sustainability, Franz is pragmatic and balances idealism with the realities of living in a bustling city. He is also hoping for science to solve some of the problems with the carbon footprint of the human lifestyle.

What is Influencing the Persona?

Trend:

The trend of incorporating more vegetables into daily diets:
Franz, and many other consumers, are becoming increasingly aware of the health and environmental benefits of plant-based eating. The carbon footprint of meat and dairy is being communicated as a significant factor in the increasing climate crisis. Franz wants to do his part in reducing his personal footprint, and has developed a growing interest in incorporating more vegetables into his family's daily meals.

FIGURE 8.3   AI-Image generated with OpenAI, ChatGPT-5 (August 2025).

Food Habits:
As passionate Franz and other consumers are about sustainability and plant-based
eating, they often struggle to consistently incorporate enough vegetables into the
family's meals, especially in a way that appeals to everyone's tastes. This trend
highlights the need for easy-to-prepare, tasty vegetable-focused solutions.

Tensions and the Motivation to Resolve them?

Why:

Functional Tension:
Franz wants to provide his family with more vegetables as part of a healthy diet, but
finds it challenging to make them tasty and appealing enough, – particularly for his
children. He enjoys cooking, but often faces time constraints during busy weekdays,
and he aims to find a balance between sustainability, health, and taste. It is a real
tension for him, and he is motivated to find practical solutions that work for his
family's everyday life.

Emotional Tension:
Franz believes deeply in the values of sustainability and healthy living, but feels a personal frustration when he can't meet these ideals. He is emotionally invested in helping his children grow up with a deep appreciation for nutritious, sustainable food. It is more than just a dietary goal, but a reflection of his identity and values as a parent.

When are the Tensions Most Prevalent and Relevant?

Context:

Morning and Mid-morning:
Franz envisions starting the day with a breakfast that is simple to prepare and appealing to the whole family. However, busy mornings often lead to not having breakfast for him and his family. They often buy sandwiches, protein drinks, or snack bars on their way. Franz is not aware, but his children sometimes even choose an energy drink instead.

N.B. If the project scope were broader, this trend and tensions could include other meal occasions. If the company were producing a wider range of categories, such as plant-based meat alternatives, sauces, spice mixes, and more, there could be significant potential for other meal occasions like lunch and dinner. As an example, Franz enjoys cooking plant-based meals for the family, but he faces difficulty in convincing his children to have less meat and dairy in their dinner. They love a traditional lasagna with beef and Mornay sauce, and they find Franz's attempt to make it into a vegan version horrible. A big potential in solving that for Franz!

Jobs to be Done

JTBD's for Family Franz:

Functional:

- Fast morning nutrition.
- Easy to prepare.

Emotional:

- Feeling that he is helping to mitigate the climate crisis, and is wanting to make a difference.
- Feeling that incorporating more vegetables into the family's daily diet is his way of showing care and promoting healthier eating.

Social:

- Low carbon footprint signals to the surroundings.

TABLE 8.2

Product	Why (For the Children)	Barrier (For Franz)
Ham and cheese sandwich	Easy to buy at the bakery Delicious and filling	Meat and cheese have a high carbon footprint Could be healthier …
Protein Drink	Fast and convenient Sweet and tasty	It is too candy-like in the flavors to feel good about it as breakfast
Fruit Smoothie	Fast and convenient Sweet and tasty	It has a high carbohydrate content that affects the blood sugar balance and is not very filling

Tasty:

- Taste is very important to Franz. Making vegetables tasty is a key priority due to one essential obstacle: If the taste is unbalanced, the children will not accept it and put down a veto.

What is Hired Today?

Please see Table 8.2

The Essence of the Problem to Solve

I would love for my family and me to eat more veggies during the day, but struggle to make them tasty enough to convince them.

How Might We?

How might we help Franz providing him and his family with more veggies in their diet?

What are the Potential Project Constraints?

In the innovation project "New Dawn," it is:

- **Category:** Potentially some form of smoothies in a bottle.
- **Occasion:** Starting from breakfast until lunch as the main focus.
- **Markets:** United Kingdom, Germany, and The Netherlands.

OPPORTUNITY SPACE 03: REFRESHING NOURISHMENT

Let's look at the last opportunity space for The Golden Juice Company. Again, a different target audience living in The Netherlands, but again, it could be based on consumer research across all three core markets for The Golden Juice Company.

PERSONA: YOGA YARA

Yara is 28 and an independent and driven young woman building her life and career. She lives in the Dutch countryside. Her location is peaceful, but within driving distance of Amsterdam. She and her partner, Penelope (34), have a small boy, Noah (1). Yara is the owner and instructor at a larger yoga studio that she runs out of their renovated countryside home. She offers yoga classes at the studio, but also online sessions. Yara embraces a minimalist, yet modern lifestyle, balancing her love for nature with her tech-savviness. She uses technology to expand her reach, running a well-curated Instagram page and a small e-commerce shop selling eco-friendly yoga gear. Her mornings start with personal yoga practice, followed by teaching classes. In her free time, she enjoys spending time with Penelope and Noah, but also hiking or cycling in the countryside, or catching up on her favorite wellness podcasts. Yara strongly believes in holistic health, but doesn't preach perfection. She encourages balance, knowing life can be unpredictable. She seeks quality, sustainability, and aesthetics in her purchases.

What is Influencing the Persona?

Trend:

The trend of shifting eating patterns:
Busy lifestyles have prompted a shift in eating patterns and behaviors, with more people preferring snacking and "on the go" eating habits rather than traditional "sit-down" meals. This trend has fueled the demand for portable, handheld, and single-serve food options that can be consumed quickly and conveniently, even while on the move. For Yara, this aligns with the need for nourishment between teaching yoga classes and her active lifestyle.

Food habits:
Yara prioritizes holistic health and sustainable living, but she faces challenges when it comes to finding time for food preparation. After teaching yoga classes, she seeks refreshment and nourishment, but doesn't want to compromise on quality or sustainability. Her choices reflect her minimalist and wellness-oriented lifestyle, aiming for natural, balanced options.

Tensions and the Motivation to Resolve them

Why:

Functional Tension:
Yara struggles with finding the time and energy to prepare nourishing snacks or meals after her classes. As a yoga instructor, her schedule is segmented into short breaks, making it difficult to dedicate time to meal preparation or clean up, without feeling rushed. This tension impacts her ability to maintain the particular healthy lifestyle she teaches. She is super motivated to find quick, nourishing solutions that fit into her rhythm.

Emotional Tension:
Yara values her role as a wellness guide and a mother, wanting to set an example of balance and self-care for her clients and family. The emotional tension arises from her desire to live up to those values while navigating the unpredictability and demands of daily life. It leaves her feeling stretched between intention and reality.

FIGURE 8.4   AI-Image generated with OpenAI, ChatGPT-5 (August 2025).

When are the Tensions Most Prevalent and Relevant?

Context:

Morning:
Yara starts her day with a personal yoga practice and moves directly into teaching classes. Preparing food is not her focus during this time, as she's occupied with setting up her studio and engaging with clients.

Post-Class Snack:
After her classes, Yara often feels drained and in need of a quick, refreshing boost. However, she doesn't want to spend time preparing or cleaning up, as she needs to remain presentable and energized for her next session.

Lunch or Afternoon Break:
Between her classes and managing her e-commerce shop, Yara often grabs whatever is quick and convenient. However, she prefers natural, nourishing options that align with her values of holistic health and sustainability.

Jobs to be Done

JTBD's for Yoga Yara:

Functional:

- Targeted nutrition.
- Easy to prepare.
- Health benefits.
- Refreshing.

Emotional:

- Nourishment and healthy indulgence.

Social:

- Convey holistic health.

Tasty:

- Yara is actually willing to compromise a little on taste. She knows that healthy products, potentially with added functional botanicals, may actually have an off taste. But like most people, she also appreciates a tasty product.

What is Hired Today?

Please see Table 8.3.

The Essence of the Problem to Solve

I like to have something refreshing and nourishing after I have taught a class, without going through the hassle of preparing it myself.

TABLE 8.3

Product	Why	Barrier
Kombucha	Very refreshing Low in sugar and packed with healthy bacteria from fermentation	Not really providing nutrition
Homemade smoothie	High content of all the good stuff. Like fruit, veggies, kefir, and botanicals	A mess to make and clean up afterwards Not always very tasty
Yoghurt drink	More nutritious and tasty	Not so refreshing, and a low amount of fruit

How Might We?

How might we help Yara get an easy refreshment and nourishment in between activities?

What are the Potential Project Constraints?

In the innovation project, "New Dawn," it is:

- **Category:** Potentially some form of smoothies in a bottle.
- **Occasion:** Starting from breakfast until lunch as the main focus.
- **Markets:** United Kingdom, Germany, and The Netherlands.

PROJECT SCOPE AND DECISION GATE 2

THE PROJECT SCOPE

The final project scope is a consolidation of the pre-scope. Based on the research conducted in the Understand phase, you should have a good understanding of the potential constellations of demand and the associated opportunities. You have put significant effort into developing the selected opportunity spaces. Now, assess them in the core team and choose the best ones. As you prepare to enter the next phases of the innovation process, define the project scope based on the answers to the **essential questions** gathered around the **area** of interest. Include how the project corresponds to the company **strategy** and potential **product category,** and associated **brand**. Update on what has been done in research and how the **timeline** is holding up. In the New Dawn project, the scope could include all three opportunity spaces, but it might also be narrowed down. The consumer group represented by the Career Carrie persona could be prevalent across all three markets and hold the greatest potential. That said, adjacent consumers like Family Franz and Yoga Yara also hold potential, and in some areas, reflect overlapping needs with each other and with Career Carrie. That potential can be backed by data from consumer research, demographic data, and the competitive landscape.

Gate 2 Meeting Pitch

Below, I will elaborate on what it could sound like at a Gate 2 meeting. (If your company doesn't have a Gate process, it should still be formulated in a project scope document to use as a guide throughout the process.) How to present it can be a combination of slides with info and mood boards displaying some of the emotional space to empathize with. The persona can also be brought to life by video recordings of the actual consumer who inspired the opportunity space. It can also be a fictitious person.

The Meeting

Welcome to the second gate meeting about the New Dawn innovation project.
 We set out to understand the breakfast occasion in our three core markets to identify consumer behaviors, the motivations behind, and the constellations of potential demand that could open up new opportunities for innovation.

What we did was the following:

- *Consumer segmentations to aim for the potential target audience.*
- *An online consumer questionnaire around smoothie usage and attitudes.*
- *Ethnographic research in Berlin, Amsterdam, and London.*
- *Mapping occasions of the target audience.*
- *Consumer engagement to discuss health, nutrition, and breakfast. (Focus group interviews).*
- *Competitor analysis of the smoothie category in the 3 markets, including recommendations for how to gain a point of difference.*

With the opportunity space "Busy Morning Fuel," we have scoped a clear target audience, represented by the persona: Career Carrie.

Her challenges of managing busy mornings and her intake of energy and nutrition hold unmet needs that could be solved by The Golden Juice Company. It sits within our strategy to help consumers to live healthier lives and we see a potential to create a new type of breakfast product, enabling just that.

We remain confident that a type of smoothie could be the solution and are excited to explore how to design it in a way that not only meets consumer needs but also offers a significant point of difference from what they currently choose. We are also still open to other solutions within the RTD category. We are still on track according to the pre-scope timeline and would like to have the approval to proceed into the Ideation phase.

Set time off for questions after the presentation, not during it, as it can disrupt the narrative. After the clarifying questions, the talk should center around the understanding of the **Holistic Consumer and Market Situation.** Understanding the basic behavior, needs, and motivations of consumers in general enables you to have empathy with them, their lives, and problems.

To reiterate, – empathy is essential for developing the ability to ask the right questions when you begin to focus on more specific problems that could be solved with products. It is viewed through the lens of **desirability**.

If desirability were the only objective of your company, many interesting products could be launched without thinking about profits. Reality, however, is that a company requires a profit to exist, and subsequently, the market situation also needs to be considered. These are the **feasibility** and **viability** lenses.

You need a holistic view of the consumers and the terms and conditions of the market but also covering what kind of values your company wants to have. That is the last two lenses. **Responsibility** and **sustainability** which interconnect with consumer desirability. As an example, vertical farming could be an answer to problems identified in consumers like Family Franz, but it might still not be the most viable business. Maybe sustainability does not fit completely with the motivations of Career Carrie either, but on the other hand, it is not a value that she would reject. Nevertheless, the sustainability and responsibility story it contributes with could hold significant value for the company and its strategy, mission, and vision.

With a clear project pitch and a thorough discussion of the project's potential, using the five lenses of innovation and company strategy, the gate steering group should be able to make an informed decision on whether to continue or to stop the project. **At The Golden Juice Company,** the decision to continue the project was unanimous, leading us into the fun and creative next phase of ideation.

It is the phase that many food and beverage companies jump to without doing the pre-work of the Understand and Define phases but I hope the granularity of details that can be provided by a proper understanding of **who** you are trying to solve **what** for and **why**, can provide the needed inspiration for companies to prioritize it, and allocate the needed resources. It is important to strongly emphasize that the time and money invested in understanding what to innovate should align with both the company's financial situation and the commercial potential of the pre-scope, to the extent that potential can be assessed at this stage. Very expensive research is not a guarantee of success in itself. I would argue that hands-on, scrappy research carried out by passionate people who genuinely put themselves in the shoes of the consumer is often more valuable than extensive reports from external research companies. These reports may not succeed in communicating the depth of understanding or in creating the essential soil and nutrition for empathy to grow. That said, this is of course not always the case! Many external agencies also do excellent work. What truly matters is tailoring the level and type of research to the company's unique context, resources, and the opportunity at stake.

The next phase is all about how to unfold the learnings and build the empathy to spark creativity and innovate solutions for real people and their real needs!

The ideation phase.

NOTES

1   I have used the same approach in the following two opportunity spaces as well.
2   *The Design Thinking Playbook: Mindful Digital Transformation of Teams, Products, Services, Businesses, and Ecosystems,* Michael Lewrick, Patrick Link & Larry Leifer, Wiley, Hoboken, NJ, 2018.

9 Ideate (Solve It)

THE CREATIVE MUSCLE (UNLOCK YOUR CREATIVITY)

Let's look at the sentence again about how to define innovation!

Innovation is when CREATIVITY enabled by empathy, is successful.

F&B innovation is the process of designing new or significantly improved products, services, processes, strategies, or business models to create value. The value should reflect a balance between building a viable business and offering desirable products that make a meaningful difference in consumers' lives. Produced and delivered sustainably and responsibly. During the Understand and Define phases, empathy was built to clearly understand what we are trying to solve and for whom. The empathy gained will be the enabler to the creative part of the innovation process. The process of coming up with creative ways to solve the challenge framed by the HMW question created in the opportunity space. Some might argue that all people are born creative but gradually they lose the ability as they grow older. At least it is a known fact that children's imagination is generally very high. Some participants in an ideation session might say that they are not very inventive, but most people still have a child's imagination and creativity inside them. To some extent, we all need to be creative to get through life. Many of the daily obstacles we meet, we solve with creativity.

DINNER EXAMPLE

Consider a simple example like cooking dinner. Suppose you're preparing lasagna for your family, following your usual recipe, when you suddenly learn that your daughter's vegetarian friend, will be joining. Now you wonder what to do ... Your goal remains the same. You are going to prepare a delicious meal that everyone can enjoy, but you have to adapt it slightly. You could decide to leave out the meat entirely, but what to replace it with? You need to evaluate what other ingredients you have available in your pantry. You can't follow the recipe anymore and you need to improvise. There are many creative choices you could make to replace the meat. The final vegetarian lasagne will be the result of your creative process. If the result turns out really delicious, it could be a new staple dish. If it doesn't turn out delicious, you can

DOI: 10.1201/9781003619352-11

157

laugh about it together at the table and maybe come up with ideas for how to improve it the next time.

Let's analyze the example, and look at how creativity and divergent vs. convergent thinking are connected! Creativity actually flourishes when we move between divergent and convergent thinking. Initially, a divergent movement is needed to explore possibilities. You need to consider different substitutes, think beyond the original recipe, and embrace flexibility. A divergent mindset supports this openness, ensuring no option is dismissed too soon. Instead of saying, *"I can't make lasagna without meat,"* you think, *"What are all the possible ways to replace the meat?"* This part of the cooking challenge is equivalent to the Ideation phase in the innovation process. As the process continues, the convergent movement takes over, focusing on refining and evaluating those possibilities. A convergent movement is to ask the following:

Which of these options will actually work the best? What will taste good and create a nice texture?

Here, the focus shifts from exploration to making decisions. You have to ensure the final dish is not just creative, but also feasible and satisfying. You are in the "Prototype phase" of your small cooking project. During the dinner, you will enter the Test phase and receive feedback from around the table. Your vegetarian guest might be polite and praise the result, and your children will most likely be brutally honest and pinpoint what could be improved. In your mind, the evaluation is running at a fast pace, collecting the feedback together with your own reflections, prepping for the next time lasagne is on the menu. Even if the need for a vegetarian version is not present, it could be that you will give your newly invented dish refinement and serve the 2.0 version. As you can see, the design thinking innovation process is present in our everyday lives as well. A sudden need to adapt the dish takes you through all the phases.

1. You **understand** the needs of your guest.
2. You **decide** on a vegetarian version of lasagna.
3. You **ideate,** based on what is available, using creativity and divergent thinking.
4. You make a decision and **select** a certain combination, based on what is feasible, using convergent thinking.
5. You start cooking and **prototype** the dish.
6. You **test** the prototype instantly at the dinner table.
7. You receive feedback, reflect, and **evaluate** the result.
8. The next time lasagne is on the menu, you might iterate the dish, and at some point, decide to **execute it** – turning the new version into a stable dish in your household.

The Ideation phase, as we are currently exploring, relies on a team's ability to unlock creativity and generate truly innovative solutions. This depends on nurturing creativity and enabling it through empathy with the personas used in the opportunity spaces at play.

In an open, exploratory phase like the Ideation phase, embracing a YES, AND? mindset is crucial. It is encouraged to state:

Hmm, YES! ... great idea! Could we also add ...?

This approach builds momentum, allowing ideas to evolve without immediate criticism. In contrast, shutting down ideas too soon with statements like the following:

"No, but ...," or "We already tried that!,"

stifles creativity before it has a chance to flourish. This resistance is especially common among employees who have spent years navigating company constraints – the convergent thinkers. While their experience is very valuable later on, it is essential to create an environment where they can see possibilities rather than just obstacles. Some find it easy to switch between the mindsets, some find it difficult.

TEAM CREATIVITY: WORK LIKE A JAZZ BAND!

The act of creativity can be performed in solitude, and many artists work well this way. Some seek feedback during the process, to get inspired or calibrated, and continue the work alone. Some artists work collaboratively, creating their art together as a team. A band of musicians, for example, might work this way. In some bands, the lead singer would be the main composer behind a song. Writing it alone and showing it to the rest of the band. She might sit down at the piano in the studio, play a rough chord progression, and sing the verse and chorus. The band starts to give feedback and builds. The pianist or guitarist of the band might add chords or suggest removing a chord. Do an alternative phrasing of the melody, other wordings, et cetera. The drummer or the bass player pitches in with rhythm ideas, and slowly the basic idea of the song evolves and grows into a piece of musical art.

Working as a team in the creative process of building new food or beverage products similar to a band can be very fruitful. The different strengths of the workshop participants can bring out the best in each individual and create a synergy that improves the product ideas. In jazz concerts, we can also become inspired by the concept of how an instrumental solo is performed and supported by the rest of the band. The concept is that all band members get a solo part where they improvise on their instrument. A jazz solo can be compared a bit to a brainstorm of sound. It is not prepared but will follow a framework based on the chord progression and a certain amount of time. The role of the rest of the band during the solo is called *comping*. When a musician is comping the rest of the band is providing harmonic support and rhythmic drive to accompany the soloist. Comping involves playing chords, chord voicings, or rhythmic patterns that complement the melody and improvisation of the soloist. Good comping requires active listening and sensitivity to the music being played. Comping musicians must listen closely to the soloist and the other band members, adapting their accompaniment to fit the musical context and direction of the performance.

Aspire to include the Jazz band mindset in the ideation process:

- Allow creativity and personal expression.
- Take turns and step back when you are not in focus.
- Listen to what is communicated and "comp" with feedback and builds.

CAN YOU MEASURE CREATIVITY?

Besides the natural creativity found in most people, you will also find individuals with a higher level of creativity. When larger companies hire new people, they have a process of testing the potential candidates. There is a high focus on cognitive abilities like problem solving, numerical reasoning, verbal reasoning, and logical thinking. There are great abilities to have in a company, but it is also really important to have individuals with strong creativity. Creative skills are particularly crucial in industries that rely heavily on innovation, like the F&B Industry. Returning to Rick Rubin the author of *The Creative Act*. In his book and in many interviews and podcasts, he conveys many valuable reflections on how to work creatively and how to work with creative people. In an interview, he reflected on how to measure creativity, stating that "it is almost as weighing smoke!" Indicating that it is pretty hard and potentially not possible. However, in the brilliant movie *Smoke*[1] (1995), a story about weighing smoke is told. Allegedly, the first Queen Elizabeth made a bet with Sir Walter Raleigh. Paul Benjamin (played by William Hurt) shares the story while talking with Auggie Wren (Harvey Keitel), and it goes something like this:

> *I admit it's strange, it's almost like weighing someone's soul, but Sir Walter was a clever guy. First, he took an unsmoked cigar. He put it on a balance and weighed it.*
>
> *Then he lit up. He smoked the cigar, carefully tapping the ashes into the balance pan. When he was finished, he put the butt into the pan, along with the ashes.*
>
> *He weighed what was there. Then he subtracted that number from the original weight of the unsmoked cigar...*
>
> *The difference was the weight of the smoke.*

I am not sure if Rick Rubin watched the movie and got inspired by the metaphor, but the point is that there will always be creative solutions and answers to a question. How to weigh smoke, as answered in the story, is creative, but it also has a fairly logical way. I believe it is worth considering if some sort of measuring creativity could be included in the hiring process for F&B innovation roles. It may be worthwhile to assess a candidate's creative thinking abilities through tasks such as brainstorming and idea generation, as part of a holistic profile evaluation. Relevant creativity abilities might include the following:

- The ability to generate a large quantity of ideas.
- The ability to generate diverse and unique ideas.
- The ability to develop and build on ideas.

IDEATION WORKSHOP

To run a workshop, you will need a facilitator. The role of the facilitator is to organize all the activities during the workshop. The facilitator's job is to set up a framework that supports the ability to creatively translate the outcome of the Define phase into tangible ideas to be built on. You can set up a workshop in many ways, but it is essentially around one to three days that you need to dedicate to a group of people, to come up with ideas that solve the tension defined in the opportunity spaces. First of all, you need to consider who to invite as participants. It is super important to get people with diverse mindsets and backgrounds. People who will be able to see a problem from different lenses and come up with solutions that could solve it in a variety of ways.

Depending on the size of the company and the resources you have available, it should be a good mix of the following:

- Facilitator
- Product Designer (If available)
- Product Developer
- Brand and/or Innovation manager
- Key Account Manager
- Scientist (Microbiology, Nutrition, Chemistry, Sensory, Consumer, et cetera ...)
- Artwork designers, or Sketch Artists
- Process Specialist (People who know how to run the production equipment)
- Procurement
- "Jokers" (One or two people from outside the industry that you know have a creative mindset)

Remember to include the factor of divergent vs. convergent mindsets in the mix. Normally, there is no need for personality tests to assess this. If you have worked together for some time, the dominant mindset is usually quite clear.

N.B. However, personality tests can be helpful when assembling teams for long-term collaboration and are worth considering if you need additional insights into a team's dynamics and individual complementary strengths.

In an ideation session, it will be great to have between 6 to 14 participants. If there are more people, you will need too much time to go through all the ideas. So that is not recommended.

The overarching agenda could be structured like this:

- Set the scene (Why are we here)
- Front loading (Outcome of the Define phase)
- Making sense (Discussion and group work to understand and create empathy)
- Ideation Session (Let's get creative!)
- Pitch, get feedback and vote (The outcome)

But before we go into the details of how to do the setup of the activities, there are some things to consider around how you meet and interact with each other.

Setup – Analog vs. Digital?

Digital 100%

Before COVID-19, it was a very common practice to meet and have ideation workshops in real life. When the reality of not being able to be physically present together set in, the digital solutions to meet and work together flourished. The world discovered that it was possible with the online tools available in 2020 to have good and prosperous results. In some cases, the online solutions actually came with some advantages. Digital tools enhance the efficiency of ideation workshops by instantly capturing all documentation, thoughts, and ideas, which can then easily be sorted and viewed in various ways. Fast overviews are a great foundation for iteration. The sessions can be recorded, transcribed, and summarized for later reference, which is super valuable in the work to be done in the Selection phase. Since 2020, the technology behind online meetings has improved significantly and will continue to, looking ahead. However, the advantages of online meetings don't stop here. The accessibility to participate improves as well. Participants can join from anywhere, which is very cost-effective and saves a lot of time. Time in online workshops is generally more time efficient, as the possibility for small talk is low, and hence it is easier to start on time and get the participants' attention. The biggest advantage, however, is the ability to invite a truly diverse group of people to participate in the ideation session. It is much easier to convince a busy individual from another part of the country, or even from abroad, to join an online workshop, as it better accommodates their time considerations. Of course, coordinating across time zones can be challenging, but in most cases, participants accept the rule of the majority.

Analog 100%

As mentioned earlier, ideation workshops were previously almost exclusively held in real life. Meeting face to face was the norm, and the energy of being physically present in the same space was seen as essential for sparking creativity and deep engagement. Despite the advancements in digital tools and the listed benefits from above, there are undeniable benefits to analog workshops. Being in the same room allows for spontaneous discussions and the possibility to read the other participant's body language, but also to feel the energy and purpose of the project in a way that digital platforms struggle to replicate. In that way, analog workshops naturally create a stronger sense of team spirit. The informal moments of grabbing a coffee together, and the small talk that often occur where trust is built and unexpected insights emerge. The tactile nature of writing on physical sticky notes and sticking them to a wall provides a different feeling. Handwritten notes and sketching ideas engage participants in a more hands-on and immersive way.

Another key strength of analog workshops is the natural flow of conversation – free from the constraints of muted microphones and digital raised hands. Participants can react instantly to each other's ideas, fostering a more dynamic and iterative conversation.

Hybrid

So why not take the best of both worlds and create a hybrid workshop experience? One of the key advantages of hybrid workshops is flexibility. Participants can choose to join in person or remotely according to availability, preference, and resources. However, the hybrid format does come with challenges! The creation of a seamless integration between the online and offline experience is not easily done. If not managed properly, remote participants may feel like passive observers rather than active contributors. It also creates more complexity for the facilitators to design the workshop flow and structure. Clear communication protocols are essential to keep both groups connected and engaged. The requirements for a more advanced technical setup also increase. A strong audiovisual system, intelligent cameras, and large touch screens for digital whiteboards are needed. Currently (2025), there are still issues with lag, integration, and real-time collaboration in digital whiteboards on large touch screens, but hopefully, screen producers and digital whiteboard suppliers will lift this design challenge.

When executed well and supported by seamless technical equipment, hybrid workshops offer the best of both analog and digital worlds. They provide the creative spark and human connection of physical meetings, while leveraging the inclusivity and efficiency of digital collaboration.

What to Choose?

My recommendation is to be physically present if possible, and to use digital tools to capture all the work. If that's not an option, a fully digital approach is the next best choice, as it ensures everyone has an equal voice and is easier and faster to plan. Hybrid workshops can be challenging, but when executed well, they combine the best of both worlds.

What to Use?

Use a lot of wall space to add sticky notes and put up posters with consumer insights and inspirational material. Digital, analog, or both. It will make the process better for all participants and the results more organized and usable.

Digital and analog tools to make a great workshop:

Digital tools: (Use your preferred hardware and software suppliers)

- Laptops, pads and phones
- Large touch screens
- Online meeting App
- Online Whiteboard
- Presentation App
- Image creation tools to capture sketches and create mock-ups.

AI is already being implemented into all the digital solutions above, as I write this book. What will also be particularly useful is AI Persona Bots, based on all the data from the constellations of potential demand and the identified opportunity spaces. These bots will allow participants to engage in interactive conversations with personas, prompting them to respond to questions, share insights, and explore different perspectives. By simulating realistic discussions with AI personas, participants can test assumptions, refine ideas, and potentially uncover new patterns in consumer behavior. This approach could provide a dynamic way to create empathy among participants who have not been deeply involved in the research.

Analog tools:

- Sticky Notes, blocks of paper, and large paper rolls for capturing and clustering ideas
- Writing and sketching tools
- Whiteboards and whiteboard markers
- Flip charts or a wall for group presentations
- Dot stickers for voting on ideas

BEFORE IDEATION ACTIVITIES

SET THE SCENE

A warm and calm welcome, along with a clear introduction to the workshop's purpose and ambition, is essential.

- Clearly state the objectives of the workshop. What are you trying to achieve?
- Highlight the bigger picture and how the session fits into the broader set of initiatives in the innovation project and the company strategy.
- Ensure participants understand how their contributions will be used afterward.

Additionally, it's crucial to establish ground rules:

- The facilitator manages speaking turns. When someone is speaking, others should actively listen and try to fully understand the shared input.
- Ensure a safe environment where all questions, thoughts, and ideas are valued.
- Encourage constructive, divergent feedback by fostering the "Yes, and …" mindset, rather than convergent statements.

Warm-Up Exercise:

Before you really get started, it is great fun and can loosen up the atmosphere if you warm up with an icebreaker or a short creative exercise. An activity to build momentum and get people comfortable sharing. If the participants are not very familiar with each other, it is mandatory to start with an ice-breaker. Ask simple questions like:

- If you were a pair of shoes, which one would you be?
- What's one thing people might be surprised to learn about you?

Or a short creative exercise to energize the group and start up the collaborative mindset and imaginative thinking. Be mindful not to use too much time on it, though.

FRONT LOADING

A week before the workshop, send out a pre-read, covering the project scope, its fit with the company strategy, and parts of the opportunity spaces. Keep it concise, but ensure that participants familiarize themselves with the vocabulary, personas, and some of their key tensions. It is the job of the core team to present the project scope of the workshop. First of all, present the opportunity spaces, scoped in the Define phase as the foundation for the workshop.

1. Who are we solving for?
2. What is influencing the persona?
3. What are the tensions to solve?
4. When are the tensions most prevalent and relevant?
5. What jobs are to be done by the product or service?
6. What is hired today? Why is it hired and what barriers might it have?
7. What is the essence of the problem to solve, in an insight sentence?
8. How can the insights be phrased as a "how might we" (HMW) question?
9. What are the potential project constraints?

Support the narrative with visual storytelling by showing short video clips from consumer video diaries or shop-alongs to bring consumer needs and motivations to life. Back them up with extracts from the research to support how the spaces of opportunity were created.

- Show the behavioral trends identified.
- Show an occasion journey.
- Show highlights from the competitor analysis.
- Show the F&B trend curve.
- Show the different constellations of demand selected.

PROCESSING THE INFORMATION

Before jumping into the ideation session, it is essential to ensure that participants process and internalize the information provided. Allocate some time for group work: The facilitator can plan this group work in many ways, depending on personal preference and available technology. The important goal is to immerse participants in the opportunity space(s) and start building empathy and understanding of the target audience and the constellation of potential demand they have. Empathy will help to generate meaningful and consumer-centric solutions and foster a shared alignment between the workshop participants to better assess the potential of all ideas during the joint effort of the day. It is vital to have good discussions in the groups and ask each other questions about the persona's potential attitudes toward the occasion and challenges during that time of day. A simple structure is to ask:

- What are their biggest concerns and frustrations?
- What do they value and prioritize?

Then ask why that is the case?

- What social norms and trends influence their behavior?
- What opinions do they hear from friends, media, or industry experts?

These discussions will be influenced by personal views and experiences from the participants. Reflecting on your own values and behavior is a natural part of processing the information into empathy. It is also very common for participants to mention knowing someone, such as a friend or family member, who is either just like this persona or the complete opposite. As touched upon earlier, the possibility of using generative AI to create a "Persona Chat Bot" is emerging on the market. Basically, feeding all the research data that you have collected into a chatbot, bringing the persona to life. Imagine having a conversation with Career Carrie and being able to ask all the questions you have! This is a very interesting new option; I am personally looking forward trying …However, even when AI bots are widely accessible and affordable, it is important to maintain group work and meaningful discussions at this stage of the process. A chatbot will not be able to provide participants with empathy, but it can be a fast and useful tool for enhancing their understanding through more detailed insights.

IDEATION SESSION

In this part of the journey, we start to diverge again, and all need to flex their creative muscles and generate a lot of ideas. Before we dive into examples of how to structure an ideation session, let's first explore what an idea actually means in the context of F&B innovation. The definition of an idea can sometimes be unclear, as F&B ideas exist at different levels of granularity.

What distinguishes a product idea from a concept?
What needs to be included?
Is a new service an idea?
Where do ideas for partnerships and new ways of generating profit sit?

Here is a guideline on how to define it:

TYPES OF IDEAS

There are basically four types of ideas:

1. Product ideas
2. Communication ideas
3. Service model ideas
4. Business model ideas

A product idea refers to a tangible food or beverage product that potentially could solve the HMW question from the opportunity space in scope. It can also be a product range idea, which is sometimes also named a *concept*. Hence, the common confusion. A product idea typically includes some key details such as the type of product, the main ingredients, nutritional features, potential benefits/claims, suggested packaging, and intended product experience.

A communication idea can take two directions:

1. Enhancing Consumer Understanding

A common situation arises when a company already has good products, but they are, for some reason, not very successful. A classic quote could be:

We already have the perfect products, but the reason why they are not more successful is because the consumer doesn't understand them!

In such cases, the idea focuses on improving communication to help consumers to better grasp the product and its benefits. While this type of idea is generally less relevant in a new product innovation project, it should definitely be passed on to marketing for inspiration.

2. Communication as a Product Enabler

The second direction involves a new product idea that heavily relies on effective communication to explain what it is and why consumers should use it. This can also include brand partnerships that enhance the product's credibility. Such enabling ideas are important to generate and capture, as they can play a crucial role in a product's success.

A service model idea refers to a creative approach to delivering, serving or engaging with consumers. Reshaping how they would normally interact with an F&B product or company. Service model innovations often leverage digital technologies, such as online ordering, subscription services, et cetera – to enhance convenience and personalization. A service model idea doesn't always involve a specific product. Instead, it focuses on improving the overall consumer experience.

A business model idea involves innovating how revenue is generated, partnerships are structured, or value is created. These ideas can reshape how an F&B business operates, driving competitive advantage by challenging traditional revenue streams and market dynamics. As with service models, a business model idea doesn't necessarily involve a specific product. Instead, it shapes the way value is delivered and captured for business growth. In this book, **a concept** is the description of a product idea, which can also be connected to a service or business model idea. Initially, it is formulated as an internal concept one-pager, pitched at the end of the ideation session to gather internal feedback. Later in the process, the concept is refined into a version suitable for consumer feedback. An external concept one-pager. More about the external version in the Prototype phase.

Concept One-pager (Internal)
Information that has to be included:

- What is the physical food or beverage product idea? Taste, texture, key ingredients, key benefits, packaging format, et cetera. If there is a communication idea, a service, or a business model connected, include it.
- How will the idea solve the HMW question from the opportunity space ideated against?
- What are the jobs to be done, and what is hired today?
- How does it meet the company constraints and feasibility?

The Perspective of the Idea

Ideas can also be viewed from the perspective of the motivation it originates from. Is it a "pull" idea coming from outside the company, or is it a "push" idea from within the company?

Outside–in, "We should solve this!"

The textbook design thinking approach involves generating ideas based on an understanding of consumer behaviors, motivations, tensions, occasions, drivers of preference, and both emotional and functional needs.

Outside–in, "They want this!"

Ideas are based on an important customer's specific request for a new product to fill a gap in their current assortment.

Outside–in, "They are doing this!"

Ideas based on identified competitor products that are gaining traction in the market and seeking to capture a share of that market.

Inside–out, "We can make this!"

Product ideas are based on existing process line capabilities, without certainty about market and consumer demand. It is important to recognize these distinctions to fully understand why an idea is created. However, it is equally important to emphasize that a successful innovation is obtainable regardless of which of these four perspectives and motivations it originates from.

TIME AND FLOW

Time

Time is essential when you are given the task to create innovative ideas in an ideation session. Often, facilitators allocate too little time for this part. Prioritization is often based on the theory that it is important to create as many ideas as possible, and a time constraint will help to achieve this. The theory is that generating a large quantity of ideas increases the likelihood of finding novel and valuable solutions. It can be great to incorporate a fast brainstorming exercise into the start of an ideation session, but don't let it stand alone. The risk is that you will end up with a nice, high number of ideas, but with low quality in terms of solving the consumer problem. It can be a valuable kick-starter, but make sure to allow sufficient time to understand the opportunity space during the ideation session, to generate meaningful solutions. Ideation isn't just about producing creative ideas rapidly! It requires reflection and empathy. By taking the time to explore the opportunity space, you create a foundation for ideas that are not only innovative but also relevant and impactful. Rushing this process risks overlooking key tensions and missing opportunities for deeper, more valuable solutions.

Flow

When you have time, flow is also obtainable. Being *in flow* refers to a mental state of complete immersion and focus on an activity, where you are fully absorbed and deeply involved in what you are doing. By concentrating on the task at hand, you experience a sense of effortlessness and fluidity in your mind. Thoughts and ideas flow smoothly and naturally. As a facilitator, creating a framework that nourishes and encourages flow is an essential task.

OLD IDEAS

Ideation session participants often have a *backpack* of good ideas from the past that they will bring to a workshop, and if they don't have enough time to move past those, it is the only ideas you will get. Many of them can, however, be really good ideas fitting the consumer tension scoped for the session, but some are also just ideas that the participants are in love with and don't fit. Ideas coming from passionate people inside your company must be taken seriously, even if they pop up out of context and without a deeper understanding of why they will be a product fitting in a certain opportunity space. I recommend using structured methods to capture these ideas, storing them for future projects, and assessing their potential fit. Additionally, recognizing and celebrating the creativity behind them fosters a culture where innovation thrives. Over time, dismissing ideas from passionate employees can weaken motivation. If employees consistently see their ideas overlooked or rejected without consideration, they may stop contributing altogether, leading to disengagement and a culture of passive execution. Building on this, I believe, Kaihan Krippendorff, in his book, *Driving Innovation from Within* (2019),[2] has some interesting points on how companies should harness the passion of their employees to drive meaningful innovation. He outlines a seven-point IN-OVATE model, which identifies key barriers that often prevent internal ideas from gaining traction.

1. **Intent:** Employees often stop contributing ideas due to initial obstacles. Take people's ideas seriously!
2. **Need:** Lack of clarity on what kind of innovations the company actually requires. No clearly communicated company or brand strategy.

3. **Options:** Fixation on a few ideas rather than exploring a portfolio of possibilities. The company's constraints, we already touched upon earlier.
4. **Value Blockers:** Resistance from leadership when innovation conflicts with the existing business model.

A great example from the F&B industry is when someone suggests innovating into a new category or storage condition.

Innovator: *"Let's move into the frozen aisle with this idea!"*
Company: *"No way! We don't have the logistics in place or the right retail buyer connections."*

5. **Act:** Bureaucracy demands proof before allowing action, making it difficult to test ideas.
6. **Team:** Siloed structures slow down innovation and discourage collaboration.
7. **Environment:** Internal politics, leadership behavior, and rigid structures hinder idea development.

Krippendorff's point is that recognizing and addressing these barriers as a company is key to unlocking the full potential of internal innovation. I agree with many of his arguments in general and encourage F&B companies to embrace and harness the value of entrepreneurship within an organization. However, it is important to highlight that this is mostly an issue in larger corporations, as startups and small companies tend to rely more on whatever input is available. Ultimately, whether big or small, it all depends on the company's culture and how innovation is managed.

Ideation

When it comes to the actual act of creating ideas, it is the facilitator's job to create a setting that enables the best possible conditions for the individual participants to be creative. I actually recommend asking the participants during the planning of the workshop if they have any preferences in relation to how they best generate ideas. Some like to sit alone and write on sticky notes, whereas others find it fruitful to work as a group and use each other as inspiration to get even more thoughts formulated. Use the information to structure the session and put together groups. If you, as facilitator, decide to have group brainstorms, provide the group with a structured way of organizing the ideas and taking turns. It might require a dedicated person to handle that. It is very important to use the jazz band mindset in this case.

Mood Board

The ideation session is built on the solid foundation of understanding the opportunity spaces and related personas, but as further inspiration, it is very nice to create visual mood boards. The boards should capture the essence of the target audience and some of the jobs to be done. You can also include something novel to the subject in the mood boards to disrupt and spark interesting new angles to the ideation process.

Brainstorm

Start with an "empty your brain exercise." All participants write down the thoughts and ideas they can think of. Group, or solo, depending on preference. Some prefer using mind maps, while others jot down their ideas on a page or sticky notes. Personally, I prefer sufficient time to ideate. Like 30 - 40 minutes, but it is also very widely used to accelerate the first part of generating ideas with exercises like the **Crazy 8 exercise.**[3]

It is a simple exercise where the participants fold a sheet of paper into eight sections and then rapidly sketch eight different ideas in eight minutes. One idea per section. The goal is to encourage quick thinking, creativity, and prioritizing quantity over perfection. The idea is that forcing participants to generate multiple ideas under time pressure, disrupts habitual thinking patterns and sparks unexpected solutions. It can also be a combination of *free* ideation time and a crazy 8. During the first part of the ideation, it can also be fruitful to disrupt the participants with a *curve ball*.

Some of the classics are:

- Reverse Thinking: Encouraging participants to consider the opposite of their usual approach.
- Random Input: Introducing an unrelated word, image, or concept to force new associations.
- Perspective Shift: Asking participants to ideate as if they were a well-known competitor, or a completely different industry player, as Airbnb or Patagonia.
- Target Audience Shift: Asking participants to ideate as if the target audience suddenly is changed to a well-know person or character like Bruce Lee, Homer Simpson, or Barbie.

Regardless of the method, the outcome should be a lot of initial ideas that could potentially solve the HMW question from the opportunity space. To be able to select which of the ideas to further refine, divide the team into small groups. Pitch as many ideas as possible and get the group feedback. Give all the ideas a maximum of one minute each to be pitched. Not a lot of time to do that, you might think, but a fast pitch is actually a good thing. If you struggle to pitch an idea and clearly describe what it is and why it solves the consumer tension, you should go back to the drawing board and sharpen the idea.

In the innovation project, New Dawn, a fast product idea pitch could look something like this:

Product Title: Vibrant Smoothie
What is the product, in a maximum of 16 words?

Very tasty Spinach & Banana smoothie.
Packed with Energy & Fibre

How does it answer the HMW question?

The vibrant smoothie can easily be consumed during commuting and at the office.
   It has a delicious, fresh taste and provides energy.

After each pitch, the rest of the group should provide fast feedback and builds – remember the Jazz-band mindset … Long discussions are not allowed! Based on the feedback and builds, select 2 - 3 of the ideas that the participants personally believe the most in, and refine the ideas into more solid ones. The selected ideas should be written into the internal concept one-pager. Group work or solo, depending on preference. **N.B.** The additional ideas not chosen, should still be captured on the wall or board, but in a "parking area" of potential ideas that didn't make the cut, in the first round.

The Internal Concept One-pager

The ideas generated don't need to be complete concept descriptions, but it will make ideas clearer to pitch, if different elements have been given thought. Provide tools to enable all participants to visualize the product idea. It can be digital drawing tools, or even a real sketch artist. At this point, the creator of the idea has received feedback and most likely reflected further on how to design and present it. How is the physical food product idea composed and how does it taste? How will the idea solve the HMW question from the opportunity space ideated against? Maybe revisit the essential questions leading to the HMW, such as:

- Who are we solving for?
- What is influencing the persona?
- What are the tensions to solve?
- When are the tensions most prevalent and relevant?
- What jobs are to be done by the product or service?
- Look at what is hired today and how this idea will have a point of difference.

Include if there is a service or business model connected – or the potential to include one? How does it meet the company constraints and feasibility? Also, add any immediate highlights or concerns regarding the last three lenses. Viability, Sustainability, and Responsibility. Let's take a look at how this could look in the New Dawn project!

Concept One Pager (Internal)

Based on Opportunity Space 1. (Busy morning fuel for Career Carrie)
(However, Family Franz and his desire to increase his own and his family's intake of vegetables, could also find this concept solving one of his needs and motivate him to try it.)

New Dawn Smoothie:

Spinach, Banana, Black Oats
Vibrant, energizing, and rich in fibre
No added sugar

More veggies and oats, than fruit content.

Added heritage TGJC Bramley Apple Juice, known for its crisp and tangy taste. Could be a range of 2 - 4 products. Different colors of veggies and ingredients provide a variety in taste. The smoothie needs to be visually appealing. The green color of the product should be visible. Maybe the bottle could be a transparent PET bottle. Not too big a size, maybe 300 ML. A product to be found in the chilled section (see Figure 9.1).

Product Sketch

How could the product look? (see Figure 9.1)

How does it answer the HMW question?
The vibrant smoothie can easily be consumed during commuting and at the office. It has a delicious, fresh taste and provides energy.

What could be the point of difference from what is hired today?
See Table 9.1.

> **Communication:** Maybe a partnership with a known vertical farm to emphasize sustainable veggies.

FIGURE 9.1   Smoothie bottle sketch.

TABLE 9.1

Product	Why	Barrier
Fruit Smoothie	Fast and convenient. Sweet and tasty.	It has a high carbohydrate content that affects blood sugar balance and is not very filling.
New Dawn Smoothie	**Fast and convenient. Not so sweet, but still tasty. Appealing colors.** **Added black oat fibre, high in Beta-glucan: Helps slow digestion, regulate blood sugar levels**	If the taste is impossible to balance, it will be a barrier. Price point can also be too high to attract a broader audience.

Service model: Existing model.

Business model: Existing model + expand sales to quick service shops and foodservice customers.

Company constraints and feasibility: Fulfills the constraints, but could require a new process technology to obtain acceptable color and shelf life.

Viability: A concern could be that it will be challenging to limit the amount of apple juice in the product, as the other ingredients come at a higher cost.

Sustainability: Good potential in sourcing of vertical farmed spinach – it should be investigated.

Responsibility: Locally produced and increased business will create more jobs.

This is just one example of how an internal concept one-pager could look. Below is another example of how a more service model-focused idea could be written for the same opportunity space. It could certainly also have been the other two opportunity spaces. Sometimes you will even find that an idea was originally written against a certain space but ends up fitting better in another opportunity space. These are simply examples intended to clarify how the process works tangibly, and will naturally vary across other F&B categories or in projects without a specific category or product constraints.

Next example, based on Opportunity Space 1. (Busy morning fuel for Career Carrie)

(Yet again, this concept could also be something for Yoga Yara and her business. It would be convenient for both her personal needs and a premium offering to offer her clients.)

Concept idea: Smoothie Vending Machine

Concept Pitch

A vending machine, connected to an app to be able to make personalized smoothies. Healthy and natural ingredients. The machine would contain three different veggies, four different fruits, two types of fibre, and a plant-based protein option.

Product Sketch
How does it answer the HMW question?
If the vending machine is placed in quick service retailers as 7-Eleven and larger company canteens, Career Carrie can order and make her morning smoothie fresher, healtier, and even personalized to her taste preference (Figure 9.2).

What could be the point of difference from what is hired today?
See Table 9.2

Communication: Collaboration with an app company that measures metabolism and gut microbiome, which also provides personalized dietary recommendations.
Service model: New service model for the company. It will require new investment in vending machines, and a logistics setup of filling and maintenance.
Business model: The new service model is also a new business model that will enable direct sales. It will provide consumer loyalty and high brand visibility.

FIGURE 9.2 Smoothie vending machine sketch.

TABLE 9.2

Product	Why	Barrier
Fruit Smoothie	Fast and convenient. Sweet and tasty.	It has a high carbohydrate content that affects blood sugar balance and is not very filling.
Smoothie Vending Machine	**Fast and convenient, and freshly made. To be tailored to fit personal preference and nutrition requirements.**	It might be that the vending machine is not available at certain locations.

Company constraints and feasibility: Only fulfills the constraints to a certain extent, as it doesn't include producing smoothie bottles at the company factory. The vending machine could be accompanied by a bottle product range as well, which would improve product availability.

Viability: There is a high concern that the return on investment will take a long time.

Sustainability: Not so high focus, as the vending machine units principally are adding more production of equipment than just selling retail bottles.

Responsibility: Maintaining and filling units could create more jobs.

As the conclusion of the workshop, the internal concept one-pagers are ready to be pitched, reflected on, and voted for by the participants.

PITCH AND VOTE

Often, an ideation session will end with a fast pitch of all the ideas generated, and a quick dot voting by the participants.

It can leave the participants with the feeling that all the efforts of understanding the consumer opportunity spaces and generating the ideas are not taken seriously. And frankly, it is exactly what it is. Typically, this part will be planned to happen at the end of a long session where both time and sufficient energy are in short supply. I recommend waiting until the next morning for the pitch and vote session, if possible, and to prioritize the needed time. Allowing sufficient time not only ensures space for explaining and reflecting on the ideas generated but it also enables a state of deep focus and effortless idea exchange, e.g., the sensation of flow. In this second round of pitches for the workshop group, provide the selected ideas 1 - 2 minutes each to be pitched. It is still pretty fast, but the same rule applies as in the first pitch. If it is hard to explain, it could be because it is an unclear idea.

Pitching reveals which ideas resonate intuitively with the audience and whether the other participants can see potential pathways for each idea. If they find it hard to grasp or align with the idea quickly, that's a valuable insight into refining or discarding the idea. It is the conversation after the pitch that is important and that particularly needs more time and facilitation. The facilitator should prompt the participants to ask questions, or even re-pitch the idea as they have understood it. Re-pitching it can spark new talks and builds that the facilitator should capture for future work. Introducing a moment for iterative refinement, where the facilitator encourages the participants to propose enhancements or even merge ideas, can also be very valuable. Not only does this help to filter the ideas, but it can also create a more collaborative environment, foster ownership and building further on the creativity among the participants. Use digital tools to record and transcribe the pitches and conversations. That will be useful later in the selection process. When all the ideas have been pitched, the next tricky part is the voting. Start with a fast-sorting exercise to create the best possible overview and identify doublets. Categorize the ideas into themes that align with the opportunity space(s). It is important to use the engagement and knowledge from all the participants who have just invested their passion and energy in the workshop. It is important for two main reasons:

1. The individuals who created the ideas, have the deepest understanding of what they are about, and how they solve the consumer need and fit the opportunity space.
2. They are most likely stakeholders in the company and further process, and their "buy-in" to what will be taken to the next phases can be very important.

After all the pitches, the builds, and good discussions, the participants should be given a few minutes to reflect on the ideas in silence. Encouraging quiet reflection can allow participants to assess ideas more thoughtfully and independently before voting, which can lead to more intentional decisions. If it is possible, create a dot voting session in a way that makes it hidden for the participants to know what the other participants are voting. It can heavily influence what people vote if they see which ideas are favored by influential voices in the company. It is recommended to facilitate unbiased personal opinions to surface first. Most digital whiteboards have that functionality built in, and hence a huge advantage to use. Beyond dot voting, consider incorporating multi-criteria voting, where participants rate the ideas based on the five lenses. Not in a high level of detail, but just to indicate an intuitive view on the level of innovation, alignment with company strategy, and the anticipated impact the ideas have. This adds more granularity and allows teams to see why certain ideas may be prioritized. Use voting, but use it consciously to provide indications on what excites the participants for the next phase, **The select phase.**

NOTES

1   Smoke, directed by Wayne Wang, written by Paul Auster, Miramax Films, 1995. Movie CLIP - The Weight of Smoke (1995) HD: www.youtube.com/watch?v=b_uXZZRpO-E
2   *Driving Innovation from Within: A Guide for Internal Entrepreneurs*, Kaihan Krippendorff, Columbia University Press, New York, 2019.
3   *Sprint: How to Solve Big Problems and Test New Ideas in Just Five Days*, J. Knapp, J. Zeratsky & B. Kowitz, Simon & Schuster, New York, 2016.

10 Select (Choose It)

WAR ROOM

At the end of the Ideation phase, we reach a peak of the divergent movement and start to converge with the pitch and voting exercises.

The outcome of the ideation workshop voting is valid inputs and should be used; however, be careful to not just use them uncritically and forget to assess all the other ideas that are created again. The risk is that the voting result will conclude that the ones with the highest score are also the best ideas, which might not be the case.

It is now time to further narrow down all the ideas created to the ones that are the most interesting to prototype and test.

We are now in the Select phase.

Plan this work to be done some days after the ideation workshop. Let the experience, ideas, discussion, and reflections sink in. No more than a week later, though, to still keep it top of mind. It is very important to take the time to evaluate all the ideas in depth after the workshop. Have the core team immerse themselves in this assessment and set aside a full day, or two half days. All core team members need to embrace a convergent mindset and find the balance between biased opinions and gut feeling: **A "kill your darling" mindset, but also an "I love it and will promote it with that energy" mindset!**

Take the time to really understand the ideas that originated from someone else.

The target audience's needs must be used as a filter to separate good ideas from the ideas that might sound nice, but don't really solve the tension in the opportunity space in a better way than what is already hired.

At this point, the ability to have an understanding and empathy with the target audience and the described space is the deciding factor behind selecting the best potential candidates to take into the Test phase.

FACILITATION

In most cases, it will make sense to continue using the facilitator from the ideation workshop in the Select phase. She will have all the knowledge from the previous phase to guide you and ask the questions you are missing. If the company is small and resources are more limited, the project leader can take the role.

DOI: 10.1201/9781003619352-12

This chapter was refined for grammar and fluency using ChatGPT-5.0.

ANALOG VS. DIGITAL

Depending on whether you want to work analog or digital, you can create an overview as a physical war room, with all the concepts printed out on a wall, or create a new infinity board in your software.

I would say that there are pros and cons to choosing either.

ANALOG

Creating a war room with all the ideas printed out or even visualized further by a sketch artist can create a good overview. If you have had a 100% analog ideation workshop, you might already have everything in a room you can use. It will feel more tangible and focused for most people. It will enhance the sense of urgency and curate the space and time for good conversations about the ideas.

The downside is that it is very time-consuming to create a very nice war room, and it is also not as flexible to change the sorting of the ideas and add builds or even new ideas. It is also quite cumbersome to collect the outcome after the selection workshop has been done. This is because of the large size formats and storage issues, and the possibility to put it up again if needed if you, for example, need to iterate and assess the ideas again with fresh eyes and new knowledge acquired in the Prototype or Test phase. The most efficient way of doing that is to take pictures and transfer all the information into a digital whiteboard.

DIGITAL

You can also create your war room digitally. If you already had the ideation workshop 100% online or as a hybrid, you will already have most of the content. This way is much more time-efficient, but it is also harder to keep the participants focused if everybody is sitting behind their own screen. It makes it even more difficult if not everyone will be able to attend physically.

If you are in the same room, you will have the best conditions for the conversations you need to have. This is not to say that a hybrid "select workshop" is not possible, it is just not ideal.

If you are all together in the war room, it is valuable to have a large screen and just one laptop for the facilitator's use. You can get big touch screens too, to enhance the experience and make it more collaborative. In the digital whiteboard setup, you will have many advantages to create different overviews and sorting exercises. Builds and comments can be easily added. If the idea originators are not present in the war room and a specific idea is unclear, you can call them, share the board, and get clarification.

After the process, everything is captured in the board and can be used for later work. The selected ideas can be copied onto a new board to prepare for the Prototype phase and the other phases to come.

NEXT STEPS

So, how do you go about selecting the most interesting ideas that you want to prototype and test?

First, start by presenting all the ideas again. The originators can present their own ideas, and for the remaining ones where the originators are not part of the core team, a mix of the team members can present them.

It is also possible to take an opposite approach, where you ensure that the team members don't present their own ideas. This way, you test how well the idea is understood. The conversations that arise in this situation can be valuable for building a shared understanding within the core team about the essence of the ideas and whether they address the defined consumer problems.

Allocating the time needed to have the pitches and conversations is again of high importance. It is where empathy for the consumer's situation and how the generated ideas can make a difference for them are built into the core team's minds. This is a common language to take into the Prototyping phase – valuable conversations and language to build the narrative and deliverables of the prototypes.

When all the ideas are covered and a deep understanding of the ideas has been achieved, it is time to do a sorting exercise.

CATEGORIZE

Based on the opportunity spaces which the ideas were created against, you have the first cluster. However, as mentioned earlier, sometimes ideas tend to fall into the "wrong" bucket and actually have a better fit in another space. So, go through all the ideas and check if they answer the "how might we" question from the intended space.

After that is sorted, go through the ideas and sort them into the four idea types:

1. Product ideas.
2. Communication ideas.
3. Service ideas.
4. Business ideas.

After that exercise, review the ideas and check for any duplicates: those that essentially share the same core proposition. During this discussion, you might also explore whether any ideas could be combined. For instance, if a service idea closely aligns with a product idea or a product idea is heavily dependent on a communication idea, they can be combined.

With that overview, sort the ideas into clusters that make sense.

In the New Dawn project, a common theme of different types of functionalities could be clusters.

1. Cluster: Ideas with a high focus on gut health.
2. Cluster: Ideas with a focus on high protein intake.
3. Cluster: Ideas with veggies as the key ingredient.
4. Cluster: Ideas with a focus on hydration and added vitamins and minerals.
5. Cluster: Ideas with a focus on mental performance.

And so on …

Adapt this according to what makes sense in the specific project that you are working on. Get a comprehensive overview and discuss what the ideas represent and their nature in detail. Consider the type of innovation they are: whether an idea is incremental or radical, or is it a small refinement of an existing concept, or does it introduce a transformative change?

THE LEVEL OF INCREMENTALITY

Another term that is widely used is the level of incrementality, or the incremental impact a product idea will have.

It may sound like the same term, but while incrementality and the spectrum of incremental vs. radical innovation are related, they actually focus on different aspects of innovation.

Incrementality in innovation refers to the degree to which a new product creates additional demand by expanding its consumer base, introducing new consumption occasions, or growing the overall market. High-incrementality innovations generate fresh demand by addressing unmet needs, appealing to new adopters, or shifting category boundaries.

In contrast, low-incrementality innovations primarily redistribute existing demand within a brand or category without significantly contributing to market growth. Does the idea have potential to expand the market and attract new consumers, or does it primarily shift existing demand within the category?

Understanding both the innovation spectrum and its incrementality helps to clarify its potential impact.

When categorizing the ideas, also make a preliminary assessment of the level of incrementality. It can be further validated later with more consumer research. If the internal concept one-pagers are completed as suggested, we will also have a preliminary view of the idea through the five lenses of innovation.

The high focus on **desirability** during the work up to now should provide a solid foundation to assess the market and consumer fit. **Feasibility** should be fairly clear at this stage and either support or challenge the "reason to believe" in a specific idea. **Viability** is harder to assess fully at this stage, but watch for red flags. The same applies to **sustainability** and **responsibility**.

The voting from the ideation workshop:

Finally, we have the voting result from the workshop participants, which are also essential inputs for selecting the ideas to take further.

DECREASE

To decrease the number of ideas before the final selection, the 2x2 matrix described in the recommended book – *The design thinking toolbox*,[1] is useful to make the levels of desirability and feasibility visible. See the illustration in Figure 10.1.

Basically, you have one axis of how well the idea solves the consumer tension – **the Consumer Value.** The other axis is how easy or hard it will be to produce for the company – **the Company fit.** A low fit means it will be hard to produce.

Filling the matrix with the ideas is a typical consensus exercise. The facilitator takes one idea at a time and places it accordingly to the team's consensus on the 2 axes.

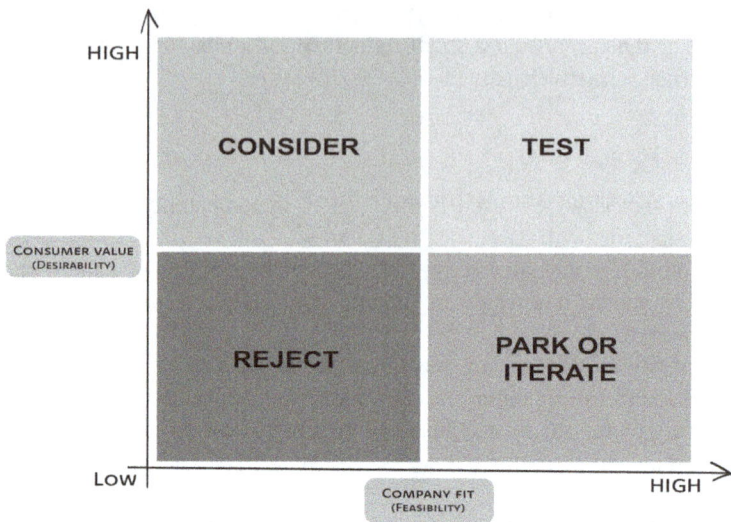

FIGURE 10.1   The 2x2 Matrix. (Re-visualized by the author, inspired by the 2x2 matrix in the *Design Thinking Toolbox.*[2])

There are four areas that an idea can end up in. The headlines indicate what to do with the ideas. They are:

1. **Reject** = The idea is not solving consumer tension very well and is further-more hard to produce with the current setup of the company.
2. **Park or Iterate** = The idea is a good fit for the company's capabilities, but it struggles to meet the consumer's needs framed in the opportunity space. The parking lot will hold ideas with good potential if they are iterated into propos-itions with a higher score on desirability.
3. **Consider** = These ideas really fit with what the consumers seem to struggle with, but are not very easy to produce or might require a new business model to work. Consider turning them into a more long-term project, looking into how to solve the feasibility issues.
4. **Test** = Bingo! The consumer will love to consume it, and the company will love to produce it.

PITCH, VOTE, AND SELECT

The final selection has to be made! What ideas to take into the Prototype phase?

The ideas located in the **top right quadrant,** called "test," are good ideas and have all the potential to be prototyped and tested. However, there will most likely be too many to take further.

In the **top left quadrant**, "consider" that there could be some really interesting ideas as well, even though the ideas would require a longer time to develop than the horizon of the project. It can be valuable to test them with consumers to validate if they find them as relevant as assessed in the matrix.

In the **bottom right quadrant,** you can either "park" the idea, to be pulled back at a later point, or in another project where the proposition may have a better consumer fit. Or, you can highlight it if it seems to have potential to easily "iterate" into a better consumer fit, before the Prototype phase.

The **bottom left quadrant** will hold the ideas to reject in this project.

Make a fast pitch round of all the ideas from "test," and the ideas from "park or iterate," and "consider." Give some time for the last reflections and conduct a voting session similar to the one from the ideation workshop.

Depending on the total number of ideas and how the votes are distributed, select 2 - 3 from "test" and 1 from "consider." If the iteration of an idea from "park or iterate" gets good support, include that as well.

The result of all the efforts and discussions has culminated in a small box of ideas. Each of them is filled with creativity, empathy, and business intelligence. Let's bring them to life in the next exciting phase, **The Prototype Phase**.

NOTES

1   *The Design Thinking Toolbox*, Michael Lewrick, Patrick Link & Larry Leifer, Wiley, Hoboken, New Jersey, 2020.
2   *The Design Thinking Toolbox*, Lewrick et al.. (Page 155–158).

11 Prototype (Build It)

In the culinary world, plating is one of the essential skills chefs are taught: the skill of combining colors and shapes to create a dish that looks irresistibly delicious. How food and beverages are served to us essentially shapes the expectations we form before tasting. "You eat with your eyes," as the saying goes.

Just as the appearance of a dish in a restaurant strongly influences our perception of its taste and overall enjoyment, retail packaging serves as the first visual impression a consumer gets of a product, whether online or in a supermarket. Later in the consumer journey, the purchased product will be used as an ingredient or consumed directly. Unpacking the product will reveal the visual appearance of what is inside. The product will be consumed, and the final evaluation of the product experience will be concluded.

In the Prototype phase, the small box of the selected ideas will need to be transformed into something tangible to show consumers and get feedback.

A prototype:
Prototyping is the area where all the topics described in the chapter about sensory and consumer science will come into play. Understanding how to design both the visual expression of the packaging, the graphical elements on the packaging, and lastly, the actual product to be consumed and enjoyed. To recap, it is about the experience of the basic tastes, flavors, how the mouthfeel, other tactile sensations, and the sound of the product all play out. However, it is also about designing the product to fit the idea that is based on all the aspects of the opportunity space and the consumer's needs.

Let's start with some definitions of what a prototype in F&B innovation can be and the level of fidelity that it can be created in.

In this book, we will use two different types of prototypes:

1. Concept Prototypes
2. Product Prototypes

DOI: 10.1201/9781003619352-13
This chapter was refined for grammar and fluency using ChatGPT-5.0.

CONCEPT PROTOTYPE

A concept prototype can be a vocal, digital, or printed version of the idea. Describing an idea in an easy, understandable way can come in different levels of fidelity.

The most basic version is the 1-minute pitch without any visual support. This version should be top of mind for all team members and able to be pitched in all relevant situations, talking with stakeholders, collaboration partners, et cetera. The digital or printed version is the external concept one-pager, as mentioned earlier, with a mockup or sketch of the packaging design. Its purpose is to present the idea to consumers at an early stage and gather their feedback about whether they understand the idea, and if it resonates with their motivations and needs.

We already went through the details of how to construct the **internal** one-pager; now let's transform it into an **external** version.

CONCEPT ONE-PAGER (EXTERNAL)

I recommend creating your own layout for the concept one-pager to fit the tone of voice of the product idea.

What to include:

The essence of the problem to solve:
In Career Carrie's case, the essence of the problem to solve was formulated like this:

> *Carrie wants to start her day feeling energized and focused, but her demanding morning routine often leaves her skipping breakfast, leading to low energy later in the day.*

Reframed to a first-person perspective:

> *I want to start my day feeling energized and focused, but my demanding morning routine often leaves me skipping breakfast, leading to low energy later in the day. I want something quick, energizing, and healthy that fits into my fast-paced routine without slowing me down.*

The intention is to check if the consumers resonate with the situation and problem to solve. The situation and narrative can also be supported by a small consumer journey cartoon of the specific occasion in scope.

Secondly, the aim is to validate whether the product proposition resonates with them as well. It can be a combination of a short marketing text and some unique selling points of the product.

The Product
New Dawn Smoothie

- Spinach, Banana, Black Oats -
  Vibrant, energizing, and rich in fibre
  No added sugar
  300 ML

Additional information:

A delicious, vibrant, and energizing smoothie designed for busy professionals to be consumed on the go.

It will come in a range of four different taste variants. To be found in the chilled section at your local retailer.

Include a picture of how the product could look: a sketch drawing or a mockup picture. The mockup design can come in many fidelities and with many levels of information:product name, brand, ingredient pictures, claims, et cetera.

There is a lot of work to be done before this part can be finalized, and we will go into the details of that shortly, but at this point, you will have to go with what is feasible to make and support the core idea in the best way.

Brand

Whether or not to include your brand in the concept prototype depends on the project scope and the answers you are looking for.

- Keep it unbranded when you want unbiased feedback on the idea itself.
- Include the brand if you want to evaluate how well the concept fits the brand's identity.

Marketing Text

New Dawn Smoothies are all packed with veggies, fruit, and fibre!

Easy to bring along and deliciously fresh – making it effortless to enjoy whether commuting or at your desk.

With no prep, no mess – just great tasting natural nutrition!

PROTOTYPING THE PACKAGING

Working with the packaging of a product prototype, it can be divided into two areas:

1. The packaging. (The type of bottle, cup, container, bag, et cetera. …)
2. The design on the packaging. (How is the "what is this product" communicated?)

Both areas are very connected and should also take their root in the foundational input from the Understand phase: the deeper understanding of the target audience, the opportunity space, and the core of the idea.

The Packaging

The packaging of food and beverages comes in all shapes and sizes. It may sound simple to choose and design a packaging for your product idea, but it is actually as complex a work as the work around the physical product that goes inside.

The packaging serves as the canvas for the graphical design, and its size, shape, and materials heavily influence what can be communicated. But before those are considered, the packaging must have key functionalities in place in several areas.

Protection

The most important role of the packaging is the protecting functionality of the product. It is the container that makes it possible to transport and store the product without it being damaged. To be able to do that, certain requirements have to be met. It all depends on the type of product and its storage conditions: Ambient, chilled, or frozen.

The packaging must protect the product from physical pressure. It also needs to keep the product fresh during the shelf life, against spoilage caused by factors such as oxygen, moisture, light, temperature fluctuation, and microbial contamination.

Size and Functionality

Along with these requirements, the packaging also needs to have the appropriate size for the occasion. Is it an "on-the-go" pack, a larger multi-serve for breakfast, or whatever usage and occasion that is intended?

It has to have a usable functionality when it comes to opening and resealing. It has to have a pleasant functionality, whether it is to be consumed directly from the pack or served/used multiple times. Together with all the considerations, the shape of the packaging also needs some thought. Will there be any shape to prompt and enhance the product expectations as well?

Storage

The packaging needs to be stackable or packable in a way that it can go on a pallet. It also needs to apply to supermarket standards to fit on shelves, chillers, and freezers.

Sustainability

It is also important to have a high focus on the sustainability of the packaging material and its recycling ability.

Consumer Appeal

Lastly, but very crucial too, the shape also needs to be appealing to the consumer – aesthetically appealing to the target audience and conveying the value proposition, Convenient to store in your home shelves, chillers, or freezers, or convenient to bring "on-the-go."

The tactile feeling of the material also contributes to the total appeal and perception of the level of quality.

Price

When all these considerations have been reflected upon, the price of the packaging needs to be assessed.

Packaging can be a very costly in the total calculation of the final product. You need to be realistic. If you are a Start-Up, you could say that you have all the possibilities of making the right choices when it comes to packaging choice. But it is rarely the case.

It can be extremely costly to invest in a packaging line, and the choices you make will define a lot of the future possibilities of product launches. In many companies with existing production lines, packaging options are essentially limited to the possibilities available. An option to get more choices is to outsource the packaging to a company that can provide better options than your own production can.

Realistic vs. Wishful Thinking

Either way you go about it, when you start to prototype your ideas, all these considerations need to be reflected in what will be shown to the consumers.

Use the five lenses tool to create clarity.

That said, it is also the role of innovation to challenge the status quo and, during the process, to listen to what the packaging tensions of the consumers are.

THE GRAPHICAL PACK DESIGN

This area is more flexible and, in some ways, even more important to get right!

One thing is that the packaging meets all the criteria from above, but if the consumer never chooses to buy the product ... well, then it really doesn't matter if the oxygen barrier in the pack is second to none. Remember the consumer's purchase journey and the importance of spotting a product in the supermarket? The way a product's design supports or even enhances expectations is crucial. The graphical design on the packaging is one of the most vital purchase triggers.

We already touched upon many of the design elements in the packaging expectations example of the Sour Cherry – Nordic Kombucha, in the Understand phase.

There are many factors to consider when designing packaging! Not only what to include, but also where to place each element. Unless the packaging type is very unconventional, four sides need to be considered. The front of the pack, also known as *the primary display panel (PDP)*,[1] is the most important area, and where most of the communication to instantly connect with the consumers happens.

Front of pack (PDP): (The usual suspects ...)

- Brand name, logo, and additional brand identity elements.
- Product name that clearly describes what the product is.
- The flavor variant that describes the main taste profile.
- The taste cues that describe what to expect of the taste experience (e.g., vibrant, tangy, refreshing, creamy, et cetera.).
- Key claims and benefits such as better health, refreshment, satiety, high content of XYZ, no added XYZ, free from XYZ, et cetera.
- Hero image – An appetizing product image or serving suggestion.
- Net weight/volume – that indicates the quantity of the product.
- Tagline or callout – that reinforces brand messaging or positioning (e.g., "Made with 100% Natural Ingredients").

The back of the pack is where consumers turn to for detailed information about what is inside and how to use the product.

Back of pack: (The usual suspects ...)

- Nutrition facts panel and ingredient list.
- Allergen warnings.
- Storage instructions about how to store the product, before and after opening.
- Usage or preparation instructions – that explain how to prepare, consume, or serve the product. (Can also be located on a side.)
- Barcode and legal information – this includes UPC/EAN code, batch number, and manufacturer details.

The two sides are where additional information and storytelling are normally located.
   The sides: (The usual suspects ...)

- Brand story – a short narrative about the brand's origins, mission, or values.
- Sustainability messaging – information about packaging recyclability, fair trade sourcing, or carbon footprint.
- Icons or "quick-to-read" symbols reinforcing product benefits.

Brand Identity

Most companies with existing brands will have the entire brand identity defined in a set of brand guidelines. They are used to provide consistent usage of how to communicate the brand identity, enabling consumers to recognize the brand even during brief exposure.

A brand guideline will include some of the following points:

Brand Mission and Values

A brief overview of the brand's mission, values, and positioning in the market.

Tagline

A tagline is a consistent, strategic phrase that represents a brand's identity, mission, or value proposition. It is typically long-term and used across multiple marketing materials to build recognition.

Some well-known F&B examples are:

Snickers: "You're Not You When You're Hungry."

Red Bull: "Red Bull Gives You Wings."

Logo Usage

Guidelines for the correct usage of the brand logo, including clear space, size, and placement.

Color Palette

Specification of the brand's color scheme, including primary and secondary colors, as well as guidelines for color usage and combinations.

Typography

Guidelines for typography, including the selection of fonts, font weights, and styles for headlines, body text, and other typographic elements.

Visual Elements and Hierarchy

Description of visual elements such as graphics, patterns, icons, or illustrations that are part of the brand's visual identity.

The visual hierarchy gives directions on how to guide the viewer's eyes through the packaging design. It highlights important information such as the brand, product name, key benefits, taste, et cetera.

Photography Style

Guidance on the style and treatment of photography or imagery used in association with the brand, including composition, tone, and subject matter.

Voice and Tone

Guidelines for the brand's voice and tone in written communication, including examples of language style and messaging principles.

Existing Brand

In prototyping the graphical design, these brand identity parameters might be set and will provide the framework for the design. It may also be the case that, from a brand identity perspective, it is considered to renew or revitalize the brand. If that's the case, it can make sense to prototype different suggestions to gather consumer feedback on which new design resonates best.

N.B. If you work in a medium or large company and the brand is very well known and established, it will, however, be advisable to make brand identity changes a separate work stream. Those types of changes can have huge implications throughout an organization and considerable costs attached.

New Brand

If you are starting a new brand together with the launch of the product(s), you have a substantially different starting point and challenge. You need to have the brand identity in focus from the beginning of the process.

In addition to the brand identity, the graphical packaging design consists of many explicit and implicit descriptions and product cues. Let's have a look at those!

The Product Name

The product needs a name that clearly describes what it is. Some products are very well known and easy to name. Like coffee, for example. If the product is not made from actual coffee beans, but a mixture of ingredients providing a coffee-like experience, then what to call it?

This can be difficult, and there is legislation to be checked before naming a product. Legislation that differs from market to market. If you have come up with a radical innovation that doesn't really have a known term, you need to be creative and come up with a catchy and relatable name.

All product names have been introduced at some point and have been adopted later on as the general term. While Coca-Cola popularized the term *cola*, it eventually became a generic name for carbonated soft drinks with similar flavors.

Colors

Colors in graphical packaging design serve both explicit and implicit purposes.

Explicitly, they can communicate direct messages about flavor, product type, or brand identity. Implicitly, they can influence consumer perception and emotions, often subconsciously shaping expectations about the taste experience and product quality. By carefully selecting colors, brands can create a cohesive visual identity, enhance consumer trust, and strengthen product positioning.

Taste

Colors supporting the taste variety can be chosen to emphasize what to expect from the taste profile and are a powerful tool in setting consumer expectations regarding the taste. For example, bright yellow and orange are commonly associated with citrus flavors, while deep red might signal berry or spicy flavors. Using color strategically can help brands guide consumers toward the right flavor choice without needing extensive text or imagery.

Beyond taste cues, colors contribute to the emotional appeal of the product. Different hues and tonalities evoke distinct emotions.

The terms **Hue** and color are, according to Klimchuk and Krasovec,[2] used interchangeably, but hue often refers to the degree of similarity between colors. A color can be a single hue like red, orange, yellow, green, blue, and violet, or combinations thereof. Hue is the attribute that differentiates one color family from another and serves as the foundation for creating various tones and color combinations in design.

Tonality or color value refers to the overall lightness or darkness of a color, influenced by the balance of shades, tints, and tones. It affects how colors interact, their visual weight, and the mood they create in design, playing a key role in brand perception and packaging aesthetics.

Let's say you're designing packaging for a new organic herbal tea brand. **Hue:** You might choose green to signal freshness and natural ingredients. **Color tonalities:** You can adjust the hue to create different moods:

- A light pastel green (tint) for a soft, delicate, and refreshing feel, great for chamomile or mint tea.
- A deep forest green (shade) for an earthy, rich, and premium look, ideal for bold herbal blends.
- A muted sage green (tone) for a calm, sophisticated, and wellness-focused aesthetic. Like a Japanese Sencha Tea.

Each tonality of green still belongs to the same hue family but conveys a different message to consumers.

Category Color

In many cases, entire product categories develop strong color associations. For example, plant-based or organic foods are often packaged in shades of green to reinforce naturalness and sustainability, while dairy products frequently use blue or white to convey freshness and purity. These industry conventions help consumers to quickly identify and categorize products on store shelves.

Color choices can also be made at the company level. I once worked for a company where all gluten-free products had to be packaged in purple. There's no inherent color logic linking purple to gluten-free, but it stood out. The idea was to make it easy for the brand-loyal consumers to identify the gluten-free products in the supermarket aisles. If the idea of using purple was inspired by another company using it for gluten-free, I am not sure, but it was evident that it inspired other companies to do the same.

Taste and Texture Visualization

A "hero image" placed in the PDP generally refers to the main, eye-catching image used to showcase a product most appealingly – often emphasizing its best features. It can be the product or ingredient image on the packaging to provide a clear representation of the product or what is the product's main ingredient.

It is also very common to display it as a serving suggestion designed to highlight the product in its most appetizing form. To avoid misleading consumers, packaging often includes a "Serving Suggestion" label to clarify that certain elements, such as garnishes or additional ingredients, are not included. While photorealistic images dominate a lot of F&B packaging, some brands choose hand-drawn sketches as illustrations to communicate artisanal, natural, or nostalgic qualities. Others choose minimalistic vector art, which can provide a clean, modern aesthetic expression.

To create strong taste and texture cues, brands use a variety of visual techniques to enhance appeal. They include:

- Close-up shots: Emphasizing textures like crispy, creamy, or juicy elements to enhance sensory appeal.
- Dramatic lighting: Using highlights and shadows to make products appear fresher, crunchier, or more indulgent.
- Ingredient breakdowns: Showcasing the individual components of the product. It could be layers of all the ingredients on top of each other.
- Dynamic motion images: Pouring shots, drizzles, or splashes to evoke freshness and indulgence.

Taste Descriptors

To enhance taste appeal, use concise, evocative statements that highlight flavor and texture. These can include descriptors like sweet, luscious, indulgent, delightful, yummy, et cetera, for flavor.

To cue the texture, it could be something like smooth, creamy, crunchy, chewy, et cetera. Taste descriptors play a crucial role in shaping consumer expectations and creating a sensory connection with the product.

Key Benefits (Claims)

In many projects, the key benefits are the most discussed topic. The unique selling points (USPs) that differentiate your innovation from the competition. The function is somewhat the same as the taste descriptors, but in this case, it creates exceptions to the product functionality instead of the taste experience.

Depending on the type of product, there can be different key benefits:

- It may be nutritional highlights such as high protein content, high in fibre, no added sugar, 0% fat, et cetera.
- Next up is what has been added like vitamins, minerals, "super food" ingredients, health oils, et cetera.
- Then the reversed situation of what has been removed like gluten-free, lactose-free, E-number-free, et cetera.

Callout Sentence

A callout sentence is typically used to highlight a key benefit, feature, or message, often to reinforce an emotional appeal or draw attention to something important. Here are some examples:

Sustainability could be highlighted with a phrase like "Better for You, Better for the Planet."

For a **Holistic health**-oriented approach, something like "Just Real Goodness" would work well.

To emphasize **Convenience**, a tagline such as "Cooking made easy" could resonate.

Heritage and Authenticity can be communicated with "A Family Recipe Since 1995."

For **indulgence**, phrases like "When you deserve that little reward" can create a sense of justification and emotional connection, encouraging consumers to treat themselves without guilt.

Legal Check

It is essential to review regulatory rules on what is legally permitted on packaging in the target launch market. Avoid including claims and statements that are clearly non-compliant. It might seem less important in the Prototype phase, as the design and statements will not be the final ones, but it can skew the quality of the consumer feedback gathered in the Test phase. It is counterproductive to know if a certain way of communicating is really resonating with consumers, if it is clear that it is not legal, and hence not possible to execute. At the same time, don't let legal constraints overly restrict early-stage communication in prototypes. Prototypes are, after all, designed to test what resonates with consumers.

It can be useful to test a range of messaging, including those that push the boundaries of current norms, as long as they are not misleading. If a particular message proves compelling to consumers, explore how it might be communicated within the legal boundaries of a specific market. There are often creative ways to achieve this.

As you can see while working with this topic, there is a fine balance between very different points of interest: a classic crash between divergent and convergent minds.

Larger companies tend to lean to the very safe side, as the financial repercussions can be substantial if the communication on the pack is not allowed. Smaller companies and startups sometimes show more agility in this space, as they can move quickly and experiment more freely during the early phases of development and smaller launches. However it is done, it is, regardless of size and type of company, important to ensure that any messaging reaching consumers must be truthful, scientifically sound, and in line with legal standards.

PACKAGING MOCKUPS

When all the elements from above have been carefully considered and there is a clear product idea to communicate to the target audience, addressing both emotional and functional needs, begin designing the visual expression of the idea as a packaging mockup.

In addition to being a visual support, the packaging mockup can also be used as the only item to show and get feedback from consumers, without the additional explanation of the external one-pager. The packaging mockup can, of course, be done in many fidelities. Choose what is feasible given the skills in your company, the resources to hire an external design agency, and the time available. It is still prototypes and they do not need to be perfect. It is, however, valuable to discuss and reflect on these communication elements. Valuable later when a potential design brief is given to a packaging design agency or an internal DTP colleague.

The packaging mockup can be sketched, printed, digital, or physical.

SKETCHED

A sketched version is a quick way to visualize the packaging idea. Working with a sketch artist, a person who can draw really well and fast in real life, brings unique benefits. A skilled sketch artist can quickly translate abstract ideas into expressive visuals, ask clarifying questions, and bring creative interpretations. This kind of collaboration can sharpen the concept early on and help teams align around a shared vision, before investing in more detailed design work. Sketched mockups are especially useful for getting fast feedback and iterating on ideas in the early stages. It is also possible to use AI-generated images, and AI tools can be particularly helpful at this stage as well. As of 2025, they do not, however, yet match the level of a good sketch artist. That said, AI is evolving rapidly and may soon be able to produce much more realistic sketches that can also be iterated and adjusted in greater detail.

PRINTED

A printed version can be a simple color print on a piece of paper. It can also be on a piece of thick carton or plastic, cut out in the packaging shape, with the ability to stand like a picture frame. This is useful for physical tests like focus group tastings.

DIGITAL

A digital version is more flexible and can be used in both physical and online tests. The packaging can furthermore be a 3D model, if seeing the packaging from many angles and even more realistically is important. It is also possible to simulate how the product would look, in both a physical shop or an online shop. As e-commerce gains more traction in F&B shopping, it will be of increasing importance to test how the products perform on a digital shelf. It does actually provide expanded possibilities to communicate more than what will fit on the physical pack.

PHYSICAL

To visualize the concept, it is also possible to create a physical packaging mockup. Going this far in the Test phase is typically not necessary to get input on the communication, and if the consumers understand what the product idea is about. It is more relevant when you want to check consumer desirability in terms of shape, size, and functionality, or perceived sustainability of the packaging.

There are two types of physical prototypes:

1. The actual type of packaging that would be feasible without any larger implications.
2. A new type of packaging that has lower feasibility and may require investments.

The prototypes can be obtained as samples from suppliers, either from their existing assortment or as newly created mockups. In some cases, they can also be 3D-printed versions. If the packaging consists of mockups or 3D prints, it will likely not be able to hold the physical product prototype. However, packaging already available in a supplier's assortment can be used with the physical product.

We will now go into the details of designing and preparing the physical **product prototype**.

PRODUCT PROTOTYPE

TASTE IS KING!

One of the differentiating aspects of fast prototyping in the food and beverage industry is the taste and functionality experience.
Normally, in design thinking, prototypes do not need to be perfect. Fast and low fidelity is what you want to get answers quickly. When the consumers are shown these low fidelities, they will be told that it is a rough idea. Not perfect, but conveying the overall idea of the concept. Be mindful that taste experiences are very much directly connected with the deep emotional triggers and subjective preferences. It is a sensitive thing to put something in your mouth, and reactions come very instinctively. Taste and smell impressions can give flashbacks to situations from the past and trigger

both good and bad memories. Similar to when you hear the song you once danced to the first time you entered the floor with your current partner or old flame.

My point is that taste experiences are very different compared to the emotional connection you might have with something like checking in at a hotel, using an app to order takeaway, or filing taxes through a government website areas where design thinking and fast prototyping are used a lot. The emotional connection to food and beverage experiences is stronger, and the frequency of interaction is much higher.

In essence, food and beverages are very taste-driven products, and everyone develops their own preferences over time based on daily experiences. Each day, we taste a variety of foods and drinks, instinctively judging whether we like them or not. Some people instantly love cheese; others don't, and some start to love it as they grow older. Some enjoy the taste of cilantro, while others strongly dislike it, and the list goes on and on … – and individual preferences change over time for most people as well.

Though we may not consciously think about it, these impressions are stored and serve as a reference for assessing new flavors. It is a well-trained skill and easily accessible when asked to provide feedback. Unless you are a trained professional, this feedback is, however, often biased, and many consumers and stakeholders may struggle to distinguish poor taste from the underlying potential of a product idea.

Tasty and Functional Prototypes

What is then needed to build suitable prototypes within F&B innovation?

Extensive knowledge about the product category in scope is needed. How to process the ingredients in a way that produces the desired product type and taste. It is essential to have skills in how to balance the taste. The knowledge of how to adjust the basic tastes into a desired direction and balance.

- How sweetness balances sourness and vice versa.
- How to adjust bitterness with sweetness and vice versa.
- How salt can also balance bitterness.
- How sour can balance salt and umami.

Knowledge of which ingredients contain aromatic compounds that complement each other, resulting in harmonious flavor combinations. Which flavor combinations and textures are preferred in the specific market? If certain functionality in the product is needed, knowledge about how to build it into the prototypes is required. Especially if it is cooking functionality.

Curry Sauce Example:

Imagine that you are working on a curry sauce for a semi-convenience dinner occasion. You have gathered a focus group of consumers who struggle with the skills to cook from scratch and with a desire to consume more vegetables. Your concept is sauces that, effortlessly, can make vegetables more interesting.

However, it is early days in the project, and the balance of the prototype taste is not very good. An assumption could be that it is not important if the taste and functionality are working completely as intended.

You could say: *"Don't mind that it is too salty and has overpowering notes of cinnamon!"* Or, *"It is not meant to be that thin. Please imagine something else!"*

At this point in the process, it is important to know if the concept and product ideas resonate with the tension they have when cooking dinner and the flavor combinations are aligned with their drivers of taste preference. Maybe they find curry sauce too polarizing and would prefer other types of sauces. They might say they need better convenient sauces, but your idea doesn't work or offer anything new. When most people have an unpleasant taste experience and are asked to give feedback on whether they like a concept of the semi-convenience curry sauce or not, they will be very influenced if the taste experience is off-putting or if the final dish is runny.

Ironically, it is not feedback on the taste experience that is key to receive at this point, but a good, balanced taste experience is an important enabler to obtain the feedback on the product proposition. It goes hand in hand.

Cooking Fish in Sulawesi Example:

Food and beverage preferences change from person to person and country to country. It is fascinating, and sometimes I have been taken by surprise in realizing that my preconceptions and assumptions were completely off. During the backpacking trip mentioned earlier, I took on a small job as a culinary consultant at a diving resort in northern Sulawesi, Indonesia. The diving resort was located on a small island and was the only building on the entire island. Surrounded by water and a lot of fish.

The chefs of the kitchen were several ladies of different ages, see Figure 11.1. The job was pre-arranged with one of the owners, but clearly not communicated to the ladies. They had no idea that I was coming or any wish to have a tall, white guy in their kitchen, telling them how to cook! To further complicate it, I don't speak Indonesian, and they didn't speak English. It was a tough start, but gradually I seemed to be more and more accepted. Communication evolved into short sentences and hands-on demonstrations of my suggestions for improving the food.

And now to my misconception. They had all this incredible fresh fish available at their front door, but I noticed that their cooking techniques to prepare them were more or less down to just one. A whole fish, deep fried! To me, it was overcooking the fish and making it very inconvenient to eat with all the bones, from head to tail! One day, I decided to show them how to filet the fish and fry it gently until firm, but still juicy and delicious. In great anticipation, I served them the fish, expecting them to be amazed by my genius preparation. However, the reaction was deeply disappointing. YES ... – they didn't like it! They actually found it undercooked and off-putting. They preferred the crunchy texture of the deep-fried version and appreciated sucking the bones clean. Lesson learned.

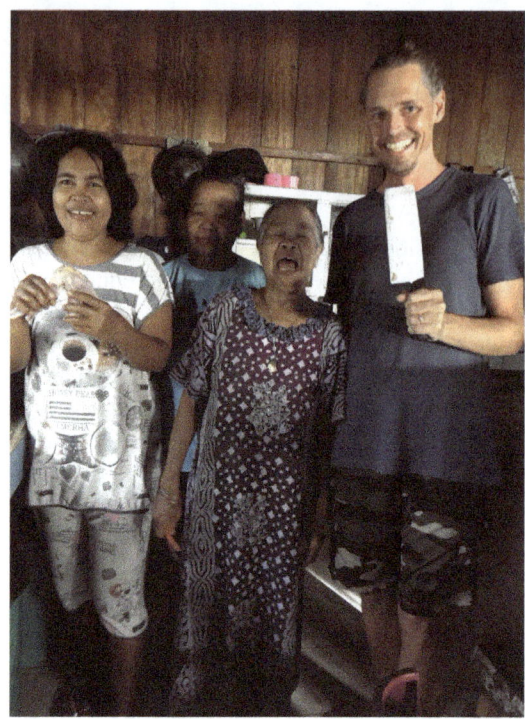

FIGURE 11.1   The Kitchen Ladies at Fadhila Cottages – Sulawesi. (Image courtesy of www. fadhilacottages.com, used with permission.)

PRODUCT PROTOTYPE TARGET BRIEF

The type of product to be prototyped must be clear. Is it a smoothie? Then it will have to fulfill the expectations of what a smoothie is. Is it a sausage? Then it will have to fulfill the expectations of what a sausage is. Is it a kombucha? Then it will have to fulfill the expectations of what a kombucha is. And so on …

In some cases, it could be that the idea is a fusion of different types of products. A new type of hybrid. That is perfectly fine. As long as you are sharp on what characteristics should be used from the different products and which should not.

The brief to the product developer working on the prototypes should be quite clear. Be sure to inform about key parameters and other important information.

Let's take a deeper look at the New Dawn Smoothie example.
You need to be very specific regarding what type of smoothie experience will fit the opportunity space. To get a good overview of the different parameters, try to list them as product idea design parameters:

IDEA

Idea number: 01. Spinach, Banana, Black Oats Smoothie – Based on Opportunity Space 1. (Busy morning fuel for Career Carrie)

If needed, there can be additional information about what the product is, in this case, a smoothie: A smoothie is a thick, creamy beverage, typically made by blending fruits, vegetables, or both, along with a liquid base such as water, milk, juice, or yogurt. Smoothies often include additional ingredients like protein powder, nuts, seeds, or sweeteners, all to enhance flavor, texture, or nutritional value. They are popular as a convenient, nutrient-dense snack or meal replacement. They can be homemade, freshly made at a cafe, come out of a vending machine, or be bought ready to drink (RTD), in a bottle.

Product Type

The product is an RTD smoothie.

Base:
Bramley Apple juice, or a mix of apple variants. (Add if relevant; some products don't have a single base ingredient.)

Main ingredients:
Spinach, Banana, Black Oats

Process:
Cold pressing + High Pressure Processing (HPP)[3]

Taste:
Good balance between sweet and sour. No added sugar. The sweetness needs to be provided by the banana and apple juice.

Flavor:
Notes of apple, spinach, banana, and oats.

Appearance:
Green color with small particles.

Mouthfeel:
Fairly thick viscosity, similar to other smoothies on the market. Smooth or with small particles.

Functional:
Added oat fibre.

Packaging:
Transparent PET bottle. 300 ML.

Shelf life:
50-60 days Chilled.

To identify which flavor combinations would be most popular with consumers, additional research can be conducted to evaluate the potential reach of various options.

TABLE 11.1

Combination	Reach (%)	Frequency
Banana + Spinach	65	1.8
Mango + Kale	54	1.2
Pineapple + Spinach	54	1.4
Apple + Carrot	45	1.9
Raspberry + Beetroot	30	1.4

TURF analysis, which stands for *Total Unduplicated Reach and Frequency*, is a statistical technique that can help to evaluate the potential reach and frequency.

For example, if the New Dawn project team would like to determine which smoothie flavor has the biggest potential in the three core markets of their project, they should gather a comprehensive list of potential smoothie flavors based on current market trends, and if it is within a category with heavy competition, the competitor analysis should also inform what product flavors are top-selling. A list of interesting combinations can be tested for consumer preferences through an online survey, followed by a TURF analysis of the results. Based on the TURF analysis results, the combinations with the highest reach and frequency can be selected for prototyping and testing. A result could look like Table 11.1.

Reach = Breadth of appeal – How many people do you reach?

Frequency = Depth of appeal – How many options would each person choose?

The taste profiles of the competitor products should also be assessed by the core team and be top of mind for the product developer. Not necessarily to imitate them, but to be aware of how they perform and how the different taste profiles are positioned in relation to each other. They can be mapped out in a matrix to get an overview of the taste profiles. This mapping exercise can be called sensory consensus mapping.

SENSORY CONSENSUS MAPPING

To provide more depth into the competitor analysis, sensory consensus mapping can be a valuable technique used to obtain a collective perception of the sensory profile of competitor products. If your company already has current products in the category, they can be evaluated as well.

You will need a facilitator with sensory experience to conduct the tasting and a group of people with the ability to taste and describe what they taste. It can be people from within the company, but also external. The product developer responsible for the prototypes is, of course, essential to be involved.

Two parameters can be considered when evaluating a product to reach a consensus. These parameters can range from low to high and may include basic tastes such as sweetness, saltiness, acidity, bitterness, and umami. Additionally, they can include other sensory attributes such as flavor intensity, flavor perception, texture, mouthfeel, chili heat, and color. Essentially, any key characteristic that defines the product. The panel will taste one sample at a time and, in turn, tell about their perception of the two parameters. In addition, the facilitator should also capture other noticeable taste impressions on the flavor profile and balance of the experience. It can be anything that the panel finds important. The facilitator can use an online whiteboard to place the samples according to the consensus and note down the comments on flavor and balance.

The consensus mapping aims to identify similarities and differences in how individuals perceive and evaluate the products. It also brings more knowledge to a team about how the product experience of the competition actually is. Sometimes it might be that you see a lot of competitor products online and view them purely on great packaging design, maybe thinking too highly of the competitors' capabilities. However, tasting them will, in some cases, be reassuring if they do not live up to the expectations created. In some cases, it can be the opposite scenario, and so on.

Smoothie Example:

In an RTD smoothie tasting, for example, you could discuss the balance of sweetness and sourness on the horizontal line.

- Place the product towards the middle if the balance is harmonious.
- Place the product towards sour (left) if the balance is predominantly sour.
- Place the product towards sweet (right) if the balance is predominantly sweet.

Secondly, how the mouthfeel of the viscosity is perceived on the vertical line. Align on what the expectations of an expected mouthfeel in a smoothie are.

- Place the product in the middle if the mouthfeel aligns with what is agreed.
- Place the product towards the thin (bottom) if the mouthfeel is thinner than expected for a smoothie.
- Place the product towards the thick (top) if the mouthfeel is thicker than expected for a smoothie.

This is illustrated in Figure 11.2.

The level of individual perception is discussed within the group tasting, and the facilitator will try to reach a consensus to place each smoothie sample on the two axes. The facilitator of the tasting will provide smoothie samples for comparison in a disclosed way. It could be five competitor products available on the market.

Consensus Mapping Comments

Competitor 01

- Flavor Profile: Bright and tropical, with dominant notes of mango, passion fruit, and coconut.
- Balance: High sourness with some natural sweetness. Really refreshing.
- With a smooth mouthfeel, but fairly thick.

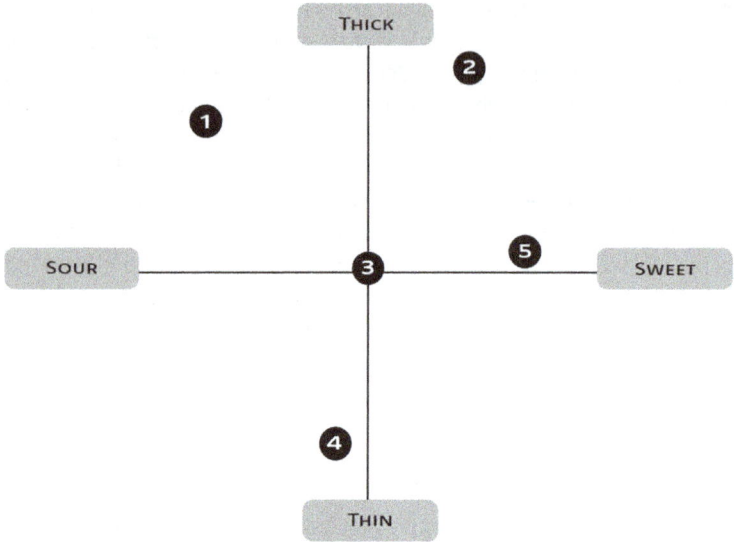

FIGURE 11.2   Smoothie Consensus Mapping.

Competitor 02

- Flavor Profile: Rich and creamy, featuring banana, peanut butter, and cocoa.
- Balance: Mildly sour with a dominant sweetness.
- A velvety texture and very thick.

Competitor 03

- Flavor Profile: High berry intensity, with dominant flavors of blueberry, raspberry, and acai.
- Balance: Well-balanced tartness and sweetness, with a vibrant and slightly tangy aftertaste.
- Very nice mouthfeel – easy to drink!

Competitor 04

- Flavor Profile: Green and earthy, blending kale, spinach, green apple, and ginger.
- Balance: Slightly tart with a subtle sweetness, complemented by an invigorating, spicy kick.
- Quite thin and watery.

Competitor 05

- Flavor Profile: Smooth and dessert-like, with notes of vanilla, almond, and caramelized banana.

- Balance: Sweet and creamy, with minimal acidity and a comforting, indulgent profile.
- Pleasant creamy mouthfeel.

After the mapping, the facilitator will reveal the products behind each sample number. This information will typically spark valuable discussions. Other sensory attributes could, of course, also be relevant. It could be color and smoothness, for example. To evaluate that, as well as create a second matrix to place that. Later in the designing the product prototype process, this map, if relevant at that stage, can guide how to iterate on the prototypes. Evaluating the product attributes of a prototype compared with competitors and the preferences of the project target audience, if available. To have an even more objective profile of important competitors, it can also be recommended to have the spiderweb profile or a PCA done on the most interesting ones. (See the Sensory and Consumer Science chapter.)

Design of Experiment (DOE)

To further validate the composition of the product parameters, it is also possible to conduct a design of experiment. It can help you identify key variables by examining the effects of multiple factors and their interactions. In a design of experiment for a smoothie, several attributes could serve as parameters to optimize the recipe. These parameters might include:

- The level of sweetness.
- Proportion of fruits (e.g., bananas, berries, mangoes).
- Proportion of vegetables (e.g., spinach, kale, carrots).
- Protein content (e.g., protein powder, yogurt).
- Fibre content (e.g., chia seeds, flax seeds).
- Texture (e.g., smoothness, thickness, grittiness).

To conduct the test, you will need to prototype the smoothie with different levels of the selected parameters. Choose only a few parameters and levels to test, as it amounts to a lot of samples very quickly. If you, for example, want to test the proportion of fruit and vegetables and the level of thickness.

Amount of banana: 20% or 30%
Amount of spinach: 5% or 10%
Level of thickness: Low or medium

This will give you eight different samples to test. As illustrated in Table 11.2, the outcome will show which combination is the most preferred within the target audience, using a liking score.

N.B. Consensus mapping, PCA, TURF, and DoE can be implemented later in the product development work as well.

TABLE 11.2

Sample	Level of banana	Level of spinach	Level of thickness	Liking
1	20%	5%	Low	5.2
2	20%	5%	Medium	5.7
3	20%	10%	Low	7.2
4	20%	10%	Medium	6.6
5	30%	5%	Low	3.4
6	30%	5%	Medium	3.5
7	30%	10%	Low	4.0
8	30%	10%	Medium	4.2

Back to the creation of product prototypes.

The product prototype can roughly be divided into three different levels of fidelity.

PRODUCT PROTOTYPE FIDELITY LEVELS

1. Low-fidelity prototype.
2. Medium-fidelity prototype.
3. High-fidelity prototype.

Low-fidelity prototype

The low-fidelity is the fastest way to create a product prototype with existing components that is still safe to taste. The components can be single ingredients, with a mix of combined ingredients, or complete products, even from the competition, that, in combination, can bring product properties and taste. They can be bought in a supermarket, without any considerations regarding sourcing or price. They are not necessarily possible to replicate in production, and hence the low-fidelity prototype is only recommended to use at an early stage and with that constraint in mind. It is a tangible prototype that consumers or stakeholders can taste and evaluate on certain explicit parameters. The aim can be top-level answers, like whether this is a concept idea that makes sense to consumers. Do they intuitively understand what it is, or do they need further explanations? It could also be more detailed answers, like what color or texture would consumers expect in a specific product type. It can, to some extent, also be about certain parts of the taste profile, like preferred flavor intensity or level of sweetness. Never get consumer feedback on the overall taste experience as if it were a real option to produce the product, though! Save that for the medium- and high-fidelity prototypes.

Furthermore, remember – low-fidelity product prototypes for F&B innovation still need to taste good and to some extent have the desired functionality! The *fast and scrappy* design thinking mindset does not apply here. **N.B.** If some of the selected ideas are a pure *me too* of a competitor's product, it can make sense to skip prototyping them. Just buy the competitor product and test it with the other prototypes. A *me too* product will most likely resonate with consumers on its core proposition.

Especially if it is a fast-growing product in retail and you have the sales data to support that observation.

It can still be a successful product to launch. Not real innovation though, more "copy-paste" ... – the work and difficulty in replicating the product, is, however, not to be underestimated. Furthermore, it also brings more company pride to innovate from scratch and not work as a "copycat."

Medium fidelity prototype

The medium-fidelity prototype needs to be able to be replicated and feasible to be scaled up to production. It is a kitchen/lab scale prototype, with mini pilot tools available to replicate the production process. All the ingredients need to be sourced through approved suppliers or at least suppliers that can be approved. Trying to use existing ingredients already listed within the company, is a good product development practice, but it is also good practice to challenge that mindset if there are good reasons to source new ingredients.

A medium-fidelity prototype can also display different variations of an idea to test. Balance the amount and types of ingredients to assess different functional solutions to create texture and taste. This fidelity of prototypes can be used to provide the same type of answers as the low-fidelity prototypes, but can also be taken to the next level of testing the real performance of the product and getting feedback on what to adjust, down to the detail.

N.B. In the New Dawn project, they will not need to create low-fidelity prototypes as they have a lot of capabilities for making medium-fidelity prototypes.

Low-fidelity prototypes can be relevant for companies and projects where the ideas are stretching current capabilities. Low-fidelity prototypes can provide valuable insights into whether an idea or project should continue or be stopped before too many resources are invested. For smaller companies, mini pilot equipment could be a semi-large investment, but an important investment to validate if a product idea actually holds a desirable, viable, and feasible potential.

High fidelity prototype

Use the high-fidelity prototype to test if the product is actually possible to scale up on a larger pilot line or the actual line that would eventually be allocated. It is quite normal that upscaling from medium-fidelity prototypes produced in smaller quantities will change the properties of the product slightly or significantly. These prototypes can be used the same way as medium-fidelity, but are normally used in the Execute phase after the product idea has been approved. This will be a validation test with the target audience to confirm that the product performs as expected and to confirm the anticipated shelf life.

PRODUCT PROTOTYPE RECIPE

When you start to produce medium-fidelity prototypes of the ideas you wish to explore the potential of, it is good practice to think ahead to the next steps after the prototype. If the concept or product idea truly resonates with consumers and you want to upscale your prototype into a larger batch of high-fidelity prototypes for further testing, it's essential to ensure proper documentation.

With that in mind, it is very valuable to have a good practice in place for the documentation of the recipes. In large established companies, recipe IT systems are standard and mandatory to use. In smaller companies, and especially startups, an easy way to create structure is to create the recipes in a spreadsheet, like in Table 11.3. It shows a rough draft of what the smoothie idea as a recipe would look like. Ensure you have a structured system to track the recipe's progression, using clear naming and numbered versions.

As mentioned earlier, it is also valuable later, when you check if the ingredients used are feasible to source from a supplier that you already use or a good potential supplier already set up to deliver larger quantities at a competitive price. The risk of using more randomly sourced ingredients is that you might not be able to replicate the taste and texture once you switch to ingredients available in larger quantities. Also, start getting an overview of the potential cost of the ingredients and packaging. Another important factor to consider is the process parameters[4] needed to produce the product and pack it into suitable packaging to ensure shelf life.

ITERATION WORKSHOP – PHYSICAL PRODUCT IDEAS

As we are working in the food and beverage industry, it makes sense to put extra effort into the physical product prototypes. If you really want to explore the sensory potential of the ideas and how different the directions could be, I would recommend having an expanded version of the prototype session. An iteration workshop centered around the physical product ideas from the Select phase. I suggest creating prototypes in low- to mid-fidelity and building on the ideas. Make iterations of the basic ideas to explore how they work in reality and how they might be improved. Additional consumer research before this workshop can provide further direction. With a TURF analysis, PCA, a DoE experiment, or specific drivers of liking, you have a really good starting point for iteration of all the sensory attributes that might be interesting to explore. Even consider broader organoleptic properties, which may be worth exploring further.

TIME

Set two days aside for this type of workshop, if it is possible. Less will also do, and in some situations, you will not have the luxury to prioritize this type of iteration workshop. It all depends on your type of company, resources available, and time constraints. In some cases, the nature of the products may require significant time to mature or settle before the result can be evaluated. It could even be products that need some sort of fermentation process to develop the taste and texture. Depending on specific factors like these, time between the two days may be needed. Fast prototyping of food and beverage ideas in a 1–2-day workshop requires skilled people and special facilities and resources.

TABLE 11.3

Recipe Name: Smoothie_Spinach_vers001

Ingredient	Weight	1,000g	Kg Price	Cost	Supplier	Article Number
Apple Juice (Cold Pressed)	500	594	1,00 €	0,59 €	XYZ	123456
Apple Juice Concentrate	20	24	3,00 €	0,07 €	XYZ	123456
Banana Pure	100	119	3,00 €	0,36 €	XYZ	123456
Spinach Pure	200	238	2,00 €	0,48 €	XYZ	123456
Lemon Juice	2	2	4,00 €	0,01 €	XYZ	123456
Black Oats (Powder)	20	24	3,00 €	0,07 €	XYZ	123456
		0		0,00 €		
		0		0,00 €		
		0		0,00 €		
Packaging, incl Label		0		0,40 €		
Production Cost		0		0,20 €		
Total Weight		1.000 g				
Total COGS				2,18 €		
Recipe Weight	842 g					
Factor to calculate 1 kg	1,19					

Process description*

Mixing (High Shear Process)
Use a high-shear mixer to ensure uniform blending of ingredients. Low speed for 5-6 minutes.

Filling into PET Bottles
Transfer the blended mixture into pre-sterilized PET 300 ML bottles using the aseptic filling line.

HPP Processing (High-Pressure Processing)
Place sealed PET bottles into HPP chambers.
Apply cold high-pressure 5,000 bar for 4 minutes.

Packaging and Storage
Apply labels and secondary packaging (e.g., cartons or shrink wrap).
Store under refrigeration (0–4°C)

Who to Invite

You need skilled chefs, product developers, and, if available, product designers to participate in the workshop. Chefs are important as they have the culinary skills to combine ingredients and preparation methods in a fast and uncomplicated way. If they are not experienced with the limitations of the f&B industry, you will also have your product developers to help steer the iterations. Product designers in the F&B industry are not that common, but still valuable to ensure consumer centricity in the ideation!

In addition to this essential hands-on experience, it will be fruitful to invite the core team and other people from within or outside the company with a "foodie" profile. Outside company people could even be first-mover consumers to help co-create the product design. It is also recommended to have a facilitator to run the day(s). Preferably, the same as in the ideation workshop, as they will know the context and hence be better able to guide the process.

Venue

To host the workshop, you need to have or find a suitable kitchen with all the possible equipment needed to create the products. It would be equipment normally available in a high-end professional kitchen. In addition, special process equipment similar to the equipment that is used in the production should be available. Some companies already have pilot equipment or even pilot departments. If there is new process technology available, it is also worth trying that out, as long as it doesn't remove the focus of the iteration task.

It can be a really good idea to write down what is possible to accomplish from a production process point of view, and what type of ingredients could be used, and which not – what is within the product design parameters, and what would be a stretch or even "out of the question." As in the war room, create a wall with all the available information for all participants to consult.

Product Design Parameters

To get a good overview of all the design parameters to consider, I recommend making an overview on a big poster and hanging it on the wall. Add the product(s) in scope and the opportunity space. To frame your specific product for this type of iteration, try to list the relevant parameters of importance. What are the options to try out, and what is out of scope?

Let's illustrate it with the smoothie example:
To recap – the smoothie idea from the ideation session had apple juice as the "base," meaning the "highest-volume" ingredient in the recipe. That is one possibility, but a smoothie could potentially also be made using other bases like other types of juices, plant or dairy liquids, or water. Also, indicate what main ingredients would be ok to add. For the sensory experience, use the relevant organoleptic properties and write what can be used to obtain a certain desired result. Lastly, in the smoothie example, there could be certain functionalities that could be interesting to obtain.

The building blocks of any product category and project scope can vary significantly. In each case, the combination of ingredients and the production process will differ. In the case of a smoothie, it could look something like Table 11.4.

WHAT TO BRING

To have room for all kinds of builds on an original idea and potentially new ideas for an entire range, a huge pantry is required. However, make sure not to use too many ingredients that are composed of multiple combined ingredients. It makes it harder to source and replicate later in the process and often more expensive. In some cases, there can be valid reasons to choose combined ingredients if a certain taste can only be achieved by their composition and process properties.

Many food and beverage factories will have a stock assortment of the ingredients already used and approved in the factory. If you can use these ingredients as your first choice, you will be popular in the procurement department. But don't let your creativity be limited either. New ingredients should be tested as well, and it is the perfect time and place to involve your B2B ingredient suppliers. Both the preferred and trusted, but also the new and aspiring.

HOW TO START

A good way to start the workshop could be with a quick "fly-in" on the project background, followed by a "deep dive" into the opportunity space(s) – especially for participants who weren't involved in the earlier phases. Thoroughly explain the basic idea and the product design parameters. After clarifying questions, divide the participants into groups if there are many.

ITERATION STARTER

To get the team started on the iteration of a physical product idea, I can suggest using the well-known exercise of Scamper.

SCAMPER

The Scamper method is widely mentioned in design thinking[5] literature. Scamper is a checklist of idea-inspiring questions based on seven prompts:

Substitute
Combine
Adapt
Modify
Put to another use
Eliminate
Reverse

Not all seven prompts may be relevant to iterate against, but just use what makes sense and try to challenge the idea in question from different angles. To convey the

TABLE 11.4

Product design parameters: New Dawn Smoothie for "Busy morning fuel"

Base	Main ingredients	Taste	Flavor	Color	Viscosity	Consistency	Functional
All fruits in a pulp or juice format	All fruits/veggies in a pulp or juice format	**Sweet:** Juice conc. Sugar or other natural sweeteners	Fruit/veg conc.	Fruit/veg conc.	Starch	Visible fibre	**Satiety:** Protein Fibre
Plant or dairy liquids	Fermented dairy products. E.g. Kefir, yoghurt	**Sour:** Juice conc. Citric acid	Natural flavors	Natural colors	Gums	Inclusion of other particles	**Immune:** Vitamins
Water	Plant-based products like nuts, seeds, dried fruit, etc.	Salt	Spices and herbs IQF		Pectin		**Other:** Botanicals Algee
Out of Scope		Artificial sweeteners	Artificial flavors	Artificial colors			

tool in a more tangible way, here are some suggestions on how it could look with the smoothie idea as an example.

S – Substitute

Think about replacing one component of a smoothie with something different. It could be replacing the liquid base. Maybe swap the apple juice base for a base of coconut water or a fermented dairy base like yoghurt or kefir. It could also be a different sweetener, replacing apple juice concentrate with honey or maybe stevia. Substitute fruits or vegetables to change sensory properties, like replacing the banana with a mango for more fruitiness.

C – Combine

To combine can be different ways of putting trends or ingredients together. Try to combine two popular beverage trends. As an example, it could be the "fruit and veggie smoothie trend" combined with the "protein shake trend." Adding concentrated whey protein to the mix of fruits and veggies. It can also take its starting point in interesting ingredients that could trigger certain emotions or claimed functionality. Combine the smoothie with functional ingredients such as vitamins or minerals. Maybe add botanicals or spices like turmeric or ginger. Combine with other types of ingredients that you could imagine that will create deliciousness in some way. Maybe dried fruit, chia seeds, pumpkin seed oil, yuzu juice, etc ...

A – Adapt

With adapt, the smoothie idea and project scope are not the best and most relevant. As an example, it could be looking at a smoothie and adapting it to another occasion or a new audience, but in this case, that is part of the foundation for the ideation. If a project is more open to exploration, maybe a smoothie could be adapted into a product for kids to have at school or as a "high-in-protein" after-fitness snack.

M – Modify (or Magnify)

With modify or magnify it is about dialing up or down certain parts of the product. (One of the principles strongly advocated in the Blue Ocean Strategy.) The sweetness level could be dialed down to bring a more "healthy" taste experience. It could also be dialing up the health benefits and supercharging the smoothie with added vitamins, collagen, or spirulina.

P – Put to Another Use

With "put to another use", this example is yet again not the best and most relevant to use. As an example, it could be looking at a smoothie and adapting it to a new purpose. Think about new contexts where the smoothie could serve a different purpose. If the project were more open to exploration, maybe a smoothie could be adapted into more of a meal. A breakfast smoothie ball with seeds and oat inclusions. Or transform the smoothie into a dessert like an ice cream or a pudding.

E – Eliminate

Eliminate is about removing components to simplify or take out redundant parts that don't bring any real value to the product. (Also, one of the principles strongly advocated in the Blue Ocean Strategy.) In the smoothie example, it could be the veggies that were removed. It could also be to completely remove the sweetness and perhaps take the smoothie into a more savory direction, maybe even adding salt. Eliminate can also be about embracing the "less is more" mindset. Perfect for brands that want to position themselves with short ingredient lists and a "clean label" communication.

R – Reverse (or Rearrange)

Reverse or Rearrange is about playing with the ratio of the product composition. The dominant taste is apple and banana, but what if it were spinach instead? It is a bit similar to dialing up or down, as described in modify.

N.B. The Scamper model is a fun way to start iteration and spark creativity. Be aware, however, that significant changes to the product might also change the type of product. The new composition might solve the consumer problem even better, or create an exciting new taste experience, but would it still be a smoothie if you change too much? Would you have to rename it, and how would that affect the proposition? If it is called a smoothie, it also has to meet the consumer's expectation of what a smoothie is. This goes for all established product types.

There can, of course, be many other ways to spark creativity beyond using a framework like Scamper. It is up to the facilitator to choose what fits the group of participants best, bearing in mind that they will most likely be very driven individuals that doesn't require, or even appreciate, much structure to explore new pathways in product design.

TASTE THE PROTOTYPES

Regardless of whether you have the iteration workshop or not, at some point, product prototypes will be ready to taste. As previously argued, you need to ensure that the prototypes are balanced in taste and convey the intended taste profile and functionality. Before showing the prototypes to consumers, it is recommended to run fast internal pre-testing with professionals who can provide actionable feedback. It is very important to use skilled tasters, such as chefs or product developers, who work with adjusting taste daily. Use them to give the product developer working with the prototype recipes valuable feedback on the product samples.

Of course, be mindful of biased opinions, but aim to foster a company culture that encourages honest feedback and open discussions about whether certain views may stem from unconscious bias. Use the existing intelligence in the company or seek feedback from outside. **Balanced taste** refers to the harmony between the basic tastes, the flavors, and the overall intensity of the flavors. It doesn't mean that the specific combination of flavors should be well-known to work, not at all. The point could be to try new combinations. When it comes to the balance of basic tastes, a specific idea might also challenge the usual ratio of sweet to sour or alter the typical level of bitterness. This is perfectly fine as well! As long as it has a purpose, and this point of difference is communicated.

The functionality could involve various aspects of the taste experience. How the texture performance provides the intended mouthfeel is the most important.

If it is a cooking product or another type of product meant to be incorporated into a final dish or drink, its intended functionality should work more or less as expected. If you, for example, are prototyping a plant-based barista product, it should not separate in the coffee. If you have granola, it should keep the crunchiness in the milk or yogurt for the appropriate amount of time. If you have a new waffle baking mix, the waffles should rise as expected and have the correct texture of a waffle – crispy on the outside and soft inside.

Mindful that it is a prototype, too much time shouldn't go into solving the issues. Don't discard the idea if it is difficult to solve quickly, but maybe wait with the taste experience and keep the prototype on paper. In cases like these, where the taste and/ or functionality provide issues, it is also important to do a feasibility check. Will it be possible to meet the expectations for the product characteristics? If the functionality is more about providing certain health benefits, it is more difficult to test that in the prototype, as the ingredients might not affect the taste. If it, for example, is a fortified product with vitamins or probiotics, it might not make sense to add anything into the prototype at this point. Once again, be mindful that adding certain micronutrients and other types of functional ingredients can have a significant effect on the taste experience.

N.B. Concept and product prototypes can also be shown to other stakeholders, but be aware that it can come with a risk to show them to stakeholders with the mandate to close or change the project. For instance, tasting a product prototype with a really unbalanced or off-putting taste can influence the trust in the project. Especially stakeholders that have the mandate, but haven't been involved in the process. It could be the CEO, sales director, board member, investors, et cetera.

It all depends on the size and type of company, of course, but try to avoid being influenced by subjective opinions from this type of stakeholder. It is not in the interest of any company.

On the positive side, it is always good to keep major stakeholders updated, if they are not directly involved in a project. It will keep them informed and included in the process and make them better equipped to make essential decisions around the project.

INVITE CONSUMERS AND GET VALUABLE FEEDBACK

If you do prioritize two days for the iteration workshop, I will recommend inviting consumers on day two for further validation of the performance of the physical prototypes. Set up a consumer interaction, like a focus group interview. Start with showing them the concept one-pager to get them into the topic. After a good discussion, let them taste the most interesting prototypes. Let them know the flavor profiles, but blind the samples.

The reason for blinding the samples is that it enables the possibility of adding competitor samples as well. It is always useful to see how they perform compared to the prototypes, without the bias of brand or product visibility. Use all the feedback to guide the next round of prototype adjustments as you move forward in the process.

How to Consume It

It is also a golden opportunity to test the packaging options with consumers and get their feedback on the experience. If it is retail products, like an RTD smoothie, you will also need to have different packaging options to test.

How to consume the product is essential for the product experience. In the "Busy Morning Fuel" opportunity space, it is an "on-the-go" type of product, but is the smoothie to be consumed from a bottle or a cup with a straw? What viscosity and texture feel pleasant to consume from, a bottle vs. a straw? In another opportunity space, it could be a 1-liter multipack, hence served in a glass. This makes the smoothies visible in another way. How does the texture and color look?

As a side note, it is common practice to buy packaging lines without any consumer insights to support the choices. The packaging line could be purchased to expand the production capacity of existing products. That is normal and needed. But packaging lines come in many shapes and forms. It is especially important to consider the possibility of different packaging formats and sizes. I recommend embracing the design thinking methodology to inform the purchase decisions of this type of equipment. Also, in the case of the production lines and their process capabilities. Getting feedback from iteration workshops can be an important step to make more future-proof investments.

SUMMARY

To summarize, there are many ways to show consumers and stakeholders the ideas incubated in a project. All are suitable at different stages in the process, for different audiences, with different purposes.

All in all, there are ten different options to show.

The Concepts and Product prototype variants and fidelity Levels:

1. One minute Pitch: Without visual.
2. Packaging Mockups: Sketched, printed, digital, or physical.
3. **Internal** Concept one-pager: With sketch artwork.
4. **External** Concept one-pager: With visual design or sketch artwork.
5. Low-fidelity product: Without pack.
6. Low-fidelity product: With pack.
7. Medium-fidelity product: Without pack.
8. Medium-fidelity product: With pack and visual design.

9. High-fidelity product: Without pack and visual design.
10. High-fidelity product: With pack and visual design.

CONCEPT PROTOTYPES

1. One minute Pitch: Without Visual

A very important prototype, in the sense that a product proposition should be easily explained. If it takes too long a speech to explain what it is and why a consumer would buy it, it can be a symptom of an idea that needs to be discarded or requires further adjustments. Or, in rare cases, just a big marketing budget to explain the wonders of this new, slightly complicated product – so when consumers finally understand what it can do for them, they will buy and enjoy it.

2. Packaging Mockups: Sketched, Printed, Digital, or Physical

There are four types of packaging mockups, and they can all come in different fidelity levels. They all aim to visualize the product idea in an easily understandable way.

1. A sketch that presents a rough idea of the product packaging design, either as a drawing or an AI-generated image.
2. A realistic image of the packaging as a printed or digital version.
3. A digital 3D model of the packaging.
4. A physical packaging mockup.

3. Internal Concept One-pager: With Sketch Artwork

The internal one-pager is the more detailed and visual pitch that can be shared internally. It states what the product idea is and includes descriptors of taste, texture, key ingredients, and packaging format. How the idea will solve the HMW question from the opportunity space it was ideated against, and what key benefits it offers.

It can also include a service or business model connected to the product idea and how it meets the company constraints and general feasibility. It is used to pitch and assess the potential of the idea across the team and stakeholders in the Evaluation phase.

4. External Concept One-pager: With Visual Design or Sketch Artwork

The external one-pager is designed to present the situation and problem to consumers and assess whether it resonates with them. A small consumer journey cartoon illustrating the specific occasion can help support the narrative and make the situation more relatable and easier to understand. The one-pager always includes what the product idea could look like: a sketch drawing, or a mockup picture. It will also include a short marketing text and some product USPs with the aim of validating if the product proposition resonates with them as well. Whether the brand is included or not depends on the the project scope and what answers you are looking for.

- Keep it unbranded when you want unbiased feedback on the idea itself.
- Include the brand if you want to evaluate how well the concept fits the brand's identity.

The external concept one-pager can also be used internally to further discuss and assess the potential of the idea.

Product Prototypes: Stand-alone or with Packaging Mockup

The product prototypes come in three fidelity levels: low, medium, high, and can all be accompanied by a pack mockup or stand alone.

5. Low Fidelity Product: Stand-alone

The low-fidelity product is a quick mix of ingredients and existing products to get a fast sense check. Does the product idea make sense, and does it seem feasible to make it tasty? Serve the prototype in a neutral cup or plate.

6. Low-fidelity Product: With Packaging Mockup

Serve the prototype in neutral retail packaging similar to what the product type is sold in today. Maybe supported by the external concept one-pager, include a packaging mockup in a sketch, image, or physical version. This setup will provide more input to base the feedback on from the consumer's point of view. Still focused on a general understanding of what the product is and the value it offers.

7. Medium Fidelity Product: Stand-alone

This prototype is perfect to serve and get consumer feedback on the taste performance. How do they find the balance of the basic tastes? The texture, consistency, and mouthfeel. The visual appearance of ingredients and colors. Sound (if relevant): The crunch, fizz, or other auditory elements while eating. It is also usable for a feasibility check. Does it seem feasible to create and upscale the process to production scale? Serve the prototype in a neutral cup or plate.

8. Medium-fidelity Product: With Packaging Mockup

When you serve the medium-fidelity product together with a realistic packaging mockup, it is easier for consumers to get an impression of how the product will look and get feedback on the initial pack shape and size, the design, and the messaging. The mockup fidelity should be close to how a real product would look. It can be a printed, digital, or physical version. The expectations created by the packaging mockup can in this way be measured against the product experience provided.

It is, in most cases, possible to serve the product prototype in a retail packaging mockup similar to what the product type is sold in today. Improve the fidelity of the packaging with a design applied to the packaging. Take inspiration from both concept one-pagers to design product visuals and the text that informs about what the product is and what the benefits are. What to expect from the taste and functionality. Communicate the core brand identity and other emotional triggers.

Only include the brand if it is important for the concept and feedback. Leave it out if it is more valuable to test the idea without branding. Print the visuals in color, sized and shaped to fit the packaging, and find a creative way to cut out and attach them. If the packaging prototype is too fragile to hold or store the actual product, serve the prototype to consumers or stakeholders in a neutral cup, plate, or undecorated packaging. Present the packaging design mockup separately alongside the serving.

9. High-fidelity Product: Stand-alone

The high-fidelity product is the ultimate feasibility test. Test whether the product is possible to scale up on a large pilot line or the actual line where it is intended to be produced. The prototype recipe needs to be realistic in terms of both the ingredients that are feasible to source and the production line process. This fidelity of product can be used for several essential purposes:

1. To test if the taste and texture are as expected. If no, go back to the recipe and adjust.
2. To test with larger groups of consumers and stakeholders whether the taste and functionality resonates with them.
3. To test the taste and texture with a trained panel to describe and compare the profile more objectively.
4. To test and compare the product against the competition in a consensus mapping or in a blind test with consumers.
5. To test the shelf life. Food safety, and sensory and visual stability.
6. To test the taste and functionality experience in a home usage test with consumers.

Serve or ship the prototype to consumers in a packaging similar to or close to the final packaging, depending on what options are available on the production line.

10. High-fidelity Product: With Packaging Mockup

This fidelity is as close to the final product as it is possible. However, in most cases, it is not possible to have prototypes with the final design directly on the packaging. If not, the product prototypes are to be supported by the packaging design mockups, printed, digital, or physical. At this point, get as close to the final design as possible.

In some cases, large pilot equipment is not available, and the only way to make a high-fidelity prototype is the actual line that the product will be produced on. The high fidelity of both the physical as well as the graphical design is normally only used past the Test phase, in the Execute phase. These tests are called validation tests and are the last resort to raise the red flag and stop a project. More about that in the Execute phase.

A GOLDEN RULE[6]:
when you serve
food or beverage PROTOTYPES
for consumers or decision-making stakeholders
it needs to TASTE GOOD
and
have the intended FUNCTIONALITY!

With prototypes ready and somewhat fine-tuned in several variations and fidelities, it is time to fully enter the next phase of testing them with consumers. This is where it gets really interesting to see if the foundation of understanding and empathy with the consumer situation and problems to solve actually align with the brilliant ideas generated and materialized through the Ideation and Prototype phases.

Did you hit it, or is it back to the drawing board?

Welcome to the next phase: **The test phase**.

NOTES

1   *Packaging Design: Successful Product Branding from Concept to Shelf*, Marianne R. Klimchuk & Sandra A. Krasovec, Wiley, Hoboken, NJ, 2012.
2   *Packaging Design,* Klimchuk & Krasovec, 2012.
3   HPP is a non-thermal (5°C – 20°C) food and beverage preservation method that guarantees food safety and achieves an increased shelf life, while maintaining the organoleptic and nutritional attributes of fresh products.
4   * The process description can include initial thoughts about the process at the medium-fidelity stage. The information will need to be more specific in the high-fidelity stage.
5   *The Design Thinking Playbook: Mindful Digital Transformation of Teams, Products, Services, Businesses, and Ecosystems*, Michael Lewrick, Patrick Link & Larry Leifer, Wiley, Hoboken, NJ, 2018. (Page 96 - originators: Osborn & Parnes).
6   No rules without exceptions, but always aspire to follow the golden taste rule. If you decide to serve prototypes that aren't tasty or don't align with the core intention of the idea, be mindful of the "why" behind your choices and be sure to have strong communication and alignment with the individuals tasting!

12 Test (Show It)

This is the phase where we fully re-enter the methodology of consumer research. This time it is about showing the consumers what have been worked on in a tangible way and be open to the feedback from them. The Test phase will determine whether the outcomes of the previous phases resonate with the consumers targeted by the identified constellations of demand and the translated potential opportunity space. If they do, it is, of course, a very rewarding feeling, but don't get too discouraged if they don't.

Several scenarios can play out in this phase. The scenario could be a state of optimism and a 'let's get it done' sense of urgency. It feels like the idea is a perfect match for an unmet need, and the potential reach, how big an audience it could attract, goes well beyond the mainstream. That is great and well done – Get cracking! However, the opposite scenario could also be the case. If the assumptions gathered in the Understand phase are not confirmed as expected, the project can enter a state of shock. The team needs to pause and reassess the direction. It can feel uncomfortable, as if all the work has been wasted. But rest assured, that it is **not** wasted work! It is valuable work that can help your company make the right decisions. An idea might not be the perfect idea yet, but the insights will offer guidance on what to adjust. They might even spark inspiration for a completely new entry point or proposition. Don't get discouraged – get cracking!

It is still possible to take what you've learned back through the waves of the innovation process and understand more, iterate more, prototype more, and test again. If time and customer commitments allow a delay, that is … This will be very project and company-dependent, but one thing is certain, it is a waste of an F&B company's resources to launch new products which are only desired by customer procurement and not by the end consumer! The customer will, of course, accept it into their assortment, but if the new product fails in solving a consumer need it will not sell and be delisted at the next window of entry and exit.

Between the two extremes, there can be lots of other scenarios, where smaller tweaks are enough to adjust the idea or a certain area in the constellation of demand needs to be replaced to create a better opportunity space.

In the phase model description, the Test phase is a separate section placed after the Prototype phase. This makes sense, as the ideas are now materialized in a way that

DOI: 10.1201/9781003619352-14

219

can be shown to consumers to get their feedback. In real life, the two phases should actually feel as one. Jump to the Test phase whenever something is ready enough, but keep working with and understanding the prototypes while waiting for the insights.

RESEARCH METHODS

Now, to the research methods that can be applied to test the assumptions from the opportunity space and the selected ideas. As mentioned in the chapter *Sensory and consumer science,* there is a strong link between these two fields of science in an F&B context. Sensory science provides structured ways to conduct organoleptic evaluation of product characteristics. Consumer science, on the other hand, captures how people actually experience and respond to those characteristics. What they like, what they expect, and whether the product delivers on its promise. Bringing the two together during testing helps to go beyond "does the consumer like it?," to "**why** they like it?" or don't like it …

It helps you not only understand whether the idea resonates, but also whether the actual product experience lives up to that idea. You can evaluate whether the sensory profile matches the expectations created by the concept and the packaging, and get insights into which specific sensory elements need adjustment to meet consumer preferences better. Like in the consumer research from the Understand phase, you have two types of approaches to consumer tests.

1. The qualitative approach.
2. The quantitative approach.

As previously mentioned, it is recommendable to conduct a combination of both types of tests. In all types of tests and their associated questions, it is essential to be clear about which key questions will provide insights into the product and concept potential. Be super mindful not to put words in the consumer's mouth or lead them to say what you're hoping to hear. Also, be aware of the bias created if the consumers know too much about your company.

At the beginning of a session, it can be beneficial to hide what the brand is, especially if it is a very well-known brand. Later in the session, the brand can be revealed, and questions can explore whether that changes anything about the consumer's attitude. The segmentation of the test participants is also a key element. You want the target audience in scope to give their opinion. So be very clear on who that is. Furthermore, it is also advisable to include a broader audience to get perspective on the proposition. Lastly, look at Everett Rogers[1]' Diffusion of Innovations Theory from the sub-chapter about segmentation and reflect on the type of adoption level that would make sense to have a focused test session with.

TESTS

There is a wide array of different tests that can be initiated. The list is long, and naturally, not all types of tests are suited for every project. Furthermore, testing involves additional costs, so it is important to carefully balance which type of research is chosen. Let's walk through key considerations, along with an overview of some of the types of tests and methodologies that can be applied and what each of them can help to inform.

WHAT ANSWERS ARE YOU LOOKING FOR?

- Do consumers understand what the product is and what it aims to solve?
- Do they recognize the demand or unmet need it addresses?
- Does the product experience live up to the expectations created by the concept and product descriptions in the external one-pager, and by the packaging design?
- Does the product meet the expectations of the target audience? Not only in terms of taste and functionality, but also of size, price, and availability?

What to ask about is one thing, but how to ask it and how to collect the answers is where it becomes a nuanced process, best handled by skilled people with consumer science expertise.

QUESTION FORMATS, ACROSS RESEARCH METHODS

There are many ways to ask questions, depending on the type of insight you are seeking, and many ways to capture the answers. Some formats are structured and scaled, others open and exploratory. Together, they shape how we understand consumer motivations, behavior, needs, and preferences.

Below are common question formats used in research[2] to gather both quantitative and qualitative data.

Yes/No Questions:

These are simple, binary questions offering only two response options. They are useful for screening and collecting clear, direct feedback.

Example:
Have you ever purchased a kombucha drink? Yes or No?

Multiple Choice:

Multiple choice questions provide a list of predefined options, helping you to gather structured and consistent responses.

Example:
Which of the following snack categories do you purchase most often?

- Chips and salty snacks
- Protein bars
- Fresh fruit
- Trail mix or dried fruit
- Other (please specify): _____

Likert Scale:

These questions measure levels of agreement, satisfaction, or frequency. They are useful for understanding consumer attitudes or perceptions.

Example:
How strongly do you agree with the following statement:

"This product met my expectations based on the packaging."

- Strongly disagree
- Disagree
- Neutral
- Agree
- Strongly agree

Semantic Differential Scale:

This format asks respondents to rate an item along a scale between two opposing adjectives, capturing emotional or sensory reactions.

Example:
How would you describe the flavor of this smoothie?
Weak ___ ___ ___ ___ ___ Bold

Ranking:

Ranking questions ask consumers to order items by preference or importance, which helps to identify what drives decision-making.

Example:
Please rank the following factors in order of importance when choosing an RTD Smoothie. (1 = most important)

- Low sugar
- Clean label (no artificial ingredients)
- Affordable price
- Innovative flavor

MaxDiff (Best-Worst Scaling):

MaxDiff is similar to ranking, but it is more effective when you need precise, scaled insights. especially across longer lists of items. MaxDiff questions ask respondents to choose the most and least important or appealing options from a set, helping to determine what truly matters.

Example:
Which of the following claims is MOST important and which is LEAST important to you when buying an RTD smoothie?

- No added sugar
- Organic ingredients
- High in protein
- Recyclable packaging

Most Important: _____

Least Important: _____

Hedonic Scale (9-Point Liking Scale):

The 9-point Hedonic Scale is a standardized method for measuring how much a consumer likes or dislikes a product or attribute. Commonly used in sensory testing, it allows you to assess expected vs. experienced liking across appearance, texture, taste, and overall impression.

Example:

How much do you like the taste of this product?

Scale from 1 (Dislike very much) to 9 (Like very much)

Common variables:

- Expectation – Appearance liking
- Experience – Taste and mouthfeel liking
- Expectation and Experience – Overall liking

Open-Ended:

Open-ended questions let consumers express their thoughts in their own words, uncovering insights that may not emerge in structured formats.

Examples:

- What is your view on functional beverages with added vitamins?
- What comes to mind when you hear the term "natural energy drink?"
- How do you decide which beverage to buy when you're looking for something healthy?

Interview Guide:

An interview guide is a flexible outline used by moderators during personal interviews or focus group discussions. It contains key themes, open-ended questions, and follow-up prompts designed to explore participants' experiences, motivations, and attitudes in depth. It could be narrative prompts that encourage participants to share small, detailed personal stories. It helps reveal emotions, context, and motivations that might not come out through direct questions.

Example:

"Tell me about the last time you tried a new snack product. What caught your eye, what did you expect, and how did it turn out?"

Unlike a fixed survey, it allows for natural conversation and probing.

Diaries / Video Ethnography:

This method asks participants to document their behaviors, choices, and experiences over time. Often, through photos, videos, or written reflections using their smartphones. It's especially useful for capturing in-the-moment consumption habits, routines, and emotional triggers in a natural setting.

Example:

Participants keep a 3-day beverage diary, recording what they consumed, when they consumed it, why they chose it, and how it made them feel, using video or voice notes and photos.

Conjoint Analysis:

Conjoint analysis is a choice-based method used to understand how consumers value different product attributes. Rather than asking directly what matters most, it presents people with sets of product combinations and asks them to choose between them. This reveals the trade-offs consumers are willing to make and identifies which features drive their decisions.

Examples:

It can, for instance, be useful to use a conjoint analysis to explore "front of pack" designs for an RTD smoothie. Each packaging picture will combine different sensory cues in a structured way. It could be:

- Two different descriptive phrases like "deliciously and creamy" or "vibrant and refreshing."
- Two different color compositions suggesting creamy or freshness.
- Two different visual elements, like a fruit picture or a fruit vector icon.

This will provide eight different combinations to show the consumer. By analyzing which combinations consumers choose most often, the research uncovers which packaging cues best shape positive sensory expectations and influence purchase intent. This is especially useful because it simulates realistic decision-making in a retail or e-commerce context. It can assist in pinpointing the specific sensory design elements that drive consumer appeal.

Implicit Methods:

Implicit methods capture non-conscious responses that people are not aware of and even less able to articulate. These techniques are often used to measure the order in which elements are noticed, the emotional reactions, and brand associations. They are particularly valuable in situations where social desirability or rational filtering might skew self-reported data.

Common tools include:

- Reaction time tasks (implicit association tests)
- Facial coding (tracking micro-expressions)
- Eye-tracking (tracking the movement and fixation of the eyes to understand visual attention)
- Biometrics (heart rate, skin conductance)

Example:
Testing consumer reactions to different packaging designs by tracking eye movement and facial expressions, rather than asking directly which one they prefer. It can also be used to measure brand bias if consumers are shown a series of logos. It could be an implicit association test to identify how different brands are unconsciously associated with naturalness or the opposite.

Consumers could be asked to quickly pair beverage brand logos (such as Coca-Cola, San Pellegrino, Red Bull, Innocent, Monster, Lipton, et cetera) with words like "natural" and "artificial."

Faster associations between a brand and "natural" suggest a subconscious alignment with purity or health, while stronger ties to "artificial" may signal image challenges.

USE EXPERTS

It's a long list, and it could be even longer. Choosing what to ask about, and how to capture it in a meaningful way, is an expertise that shouldn't be underestimated. Fast and scrappy consumer immersion advocates might look lightly on this topic, but it matters! Poorly designed questions or mismatched methods can lead to misleading insights, which can be even worse than having no insights at all, as they risk steering teams in the wrong direction. Use sensory and consumer scientists to help shape what is needed for each specific project. If the project team doesn't have that expertise in-house, collaborate with external partners who do. Be sure to choose those with proven experience in F&B sensory and consumer research.

As in all previous stages of the innovation process, the number of resources that can be allocated to the Test phase, of course, depends on the type and size of the F&B company. Startups and smaller companies will have to go with what is possible within their financial constraints; larger companies will have more choices.

ONLINE VS. PHYSICAL

Online vs. physical formats offer different benefits and some limitations, similar to the differences between digital and analog discussed in "how to set up an ideation workshop."

Online:

If you conduct testing online, you have more flexibility in screening and recruiting participants, as they don't need to travel to a specific location. They can participate from their own home, responding via video calls, recorded videos, or digital surveys. This approach makes it easier to include people from different regions or with tight schedules. It is fast and cost-effective. However, it may limit the depth of interaction. Non-verbal cues and spontaneous feedback can be harder to capture in a remote setting.

Physical:

If you conduct testing physically, you can build a stronger connection with participants. Being in the same space allows for richer, more dynamic interaction. You can observe body language, facial expressions, and subtle reactions that provide valuable insights. This is especially important in food and beverage testing, where the sensory experience plays a central role. While physical testing requires more time and coordination, it often leads to more engaged participants and deeper, more nuanced feedback.

Now, let's look at some valuable research approaches and tests worth considering to explore the potential of the prototypes:

CONCEPT TEST

Concept tests are a standard test in this phase. Offered by many established companies. A concept test helps you check if your idea actually makes sense to consumers, before you invest too much time and money. You share the **external one-pager** in an online survey and ask a potential target audience what they think. Do they get what it is? Does it solve a real need for them? Is it different enough from what's already out there?

By testing early, you get a clearer sense of whether the idea resonates, feels relevant, and has the potential to stand out. It's a way to explore if the promise you make aligns with what people are looking for, and whether they will actually consider trying it.

PACK CUES AND PRODUCT PERFORMANCE

Pack cues set the expectations, while tasting the product reveals the experience. As mentioned earlier, there are many graphical elements to consider when designing packaging. Each of these elements offers cues to the consumer about the brand, what it represents, and the anticipated product experience. These visual elements play a critical role in shaping consumer perceptions. This applies to both the brand and product identity, as well as to sensory expectations. What the product is, who is behind it, and what it might taste like. To gain insights into how consumers perceive and respond to this visual communication, let's take a look at some of the research methods that can be applied to explore it.

Conjoint analysis, as already mentioned, is a valuable tool for understanding consumer preferences and determining which elements within a range of design options are most influential or meaningful. It helps identify which visual cues consumers prioritize or associate with certain values or product qualities.

Eye tracking, when combined with qualitative interviews, can reveal implicit patterns, such as what consumers notice first, what draws their attention, and the sequence in which visual elements are processed. This method uncovers the subconscious responses that might not be captured through a conjoint analysis.

Focus group interviews are also highly valuable. By serving physical prototypes alongside multiple packaging design options, they create a setting to explore how well the visual identity aligns with the actual product experience. These sessions can highlight gaps between expectations and experience, offering valuable feedback for refining both the packaging and the product design.

HOME USE TEST (HUT)

Home Use Tests[3] are great and involve testing products in the own homes of the consumers. A more natural setting, even if it is not part of their usual routine. The method provides real-life insights into product performance, acceptance, and preferences: how the products are used in actual household environments. The main objective is to see how the consumer opens, pours, prepares (if needed), serves, and consumes the products in conditions that are not controlled. This gives insights into how consumers intuitively interact with the product. To avoid making the test situation too complex, it is recommended to have only a few prototypes to test.

Secondly, it is also interesting to see what competitor products the consumers have in their kitchens and why they have bought them. The challenge with HUTs is that they require a lot of logistical planning. Some product categories will be very easy to send out to individual homes, but other categories are more challenging. Products with a short shelf life and a need for chilled and careful handling are tricky.

Previously, the data collection was done through interviews or questionnaires, which is still a valid option. With *today's technology,* it is very valuable to have the participants answer a questionnaire while filming themselves. This will provide much more information than the answers alone, but also show the participant's body language and tone of voice, the kitchen environment, and all of these details, to create empathy with their views and situation.

STREET IMMERSIONS

Street Immersions involve taking your F&B idea or physical prototype directly to where people are. It is a fast and cost-effective way to test appeal and relevance outside traditional research settings. For example, if you want to test whether a product idea resonates with a younger audience, setting up a pop-up stand at a music festival could be a smart move. This quick and informal approach allows you to gather spontaneous feedback from a broad and diverse group. Still more segmented and relevant compared to engaging with completely random consumers on the streets or in other public areas. That can work too, but be cautious! Feedback from such interactions may reflect opinions that are too far from your target audience and could skew your insights.

CENTRAL LOCATION TEST (CLT)

A Central Location Test[4] (CLT) is a controlled research method used to evaluate consumer preferences, perceptions, and product acceptance in a local market setting. If you are a global company, it will be advisable to conduct a similar test in all the markets. There are many supporting companies offering this service, also across markets. CLTs are ideal for large-scale data collection. Especially useful for testing prototypes and comparative testing against competitors. A very efficient way to gain insights and inform the final product design brief. Later in the process, the insights can also help to inform the work on shaping the marketing strategies around the product in the Execute phase.

CLTs take place in locations where a large number of potential consumers can easily be recruited, like shopping malls or other larger venues. This setup allows researchers to gather feedback in controlled yet accessible conditions from real consumers. However, since the artificial setting differs from real-world usage conditions, this must be taken into consideration when evaluating the results. The segmentation will focus on everyday consumers representing the actual target audience, but it can also include adjacent audiences, especially if you want to explore the potential among consumers who might consider trading up or down. Participants are typically recruited on-site and screened by fit of the segment criteria. These locations should ideally include separate spaces for registration, preparation of samples, and tasting to ensure focus and consistency. The participants are guided through the test process via a specifically designed questionnaire and the help of trained professionals, which helps maintain accuracy and reliability in the results.

The products can be a combination of different prototypes, but also products from the competition. The logistical hurdles of transporting prepared samples require careful planning. The test can combine visual stimuli and physical prototypes, allowing participants to articulate their preferences or concerns about inconsistencies between visual expectations and the product experience. The primary response variable is usually consumer liking.

This would typically use a 1–9 scale:

1 = dislike very much
9 = Like very much

Variables could be:

- Expectation – Liking the appearance
- Expectation – Liking texture
- Expectation – Liking taste
- Expectation – Overall liking
- Experience – Liking the appearance
- Experience – Liking the mouthfeel
- Experience – Liking the taste
- Experience – Overall liking

Furthermore, it is also common and valuable to ask participants to what extent they would consider buying a product like the prototype. Using the same 1–9 scale provides a measurement of **purchase intent**, which is another valuable indicator of how well the product resonates with consumers. In this case, additional information on size, claims, and price can also be included to provide context.

N.B. The CLT test is also often used as a final validation test in the Execute phase. We will get back to that.

ANALYSIS

After the selected tests have been conducted, it is time to look at the feedback from the consumers. As a team, you need to take in all the information collected and assess if the product resonates with the target audience. Look at the learnings and data provided by the tests and summarize them into conclusions and implications. Score the desirability of the product idea on a simple scale.

FOUR LEVELS OF FIT

Low Fit = 25
The idea does not resonate with the target audience. Lacks relevance, appeal, or clear value. Unlikely to succeed without a fundamental rethink.

Moderate Fit = 50
There are some appealing elements, but the idea doesn't yet fully connect with user needs or desires. Potential exists, but it requires significant refinement.

Strong Fit = 75
The idea matches the expectations and interests of the target audience. It is relevant, clear, and attractive. Only minor improvements are needed.

Excellent Fit = 100
Highly compelling and clearly desirable to the target audience. The idea is well-crafted to consumer preferences and ready to move forward.

Ideas that score 25 should be discarded and maybe revisited in an iteration of the idea or turned into a new project. Add the ideas that score 50, 75, or 100 into a report, together with a final recommendation from the team. Take this report into the next phase: **The evaluation phase**.

NOTES

1   *Consumer Behavior* (Fifth edition), Zubin Sethna, Sage, New York, 2023. Page 145–146.
2   *The Consumer Insights Handbook: Unlocking Audience Research Methods*, Danielle Sarver Coombs, Rowman & Littlefield Publishers, USA, 2021.

3 *Sensory and Consumer Research in Food Product Design and Development*, Howard R. Moskowitz, Jacqueline H. Beckley & Anna V. A. Resurreccion, Wiley-Blackwell, Hoboken, NJ, 2006 (HUT: Pages 249–257).
4 *Sensory and Consumer Research*, Moskowitz et al., 2006. (CLT: Pages 241–249).

13 Evaluate (Think About It)

This is the phase where the final evaluation of the best-performing ideas takes place. There might be issues that can be fixed with smaller feasible iterations, and if so, it is okay to take an idea into this phase. Suppose there are larger issues of consumers not understanding the proposition or not finding the products relevant in general. In that case, the ideas need to be evaluated closely to assess if there is enough potential to pivot the project and go back to the stage in the process where further work on understanding and iterating the ideas towards a better fit makes sense.

INNOVATION POTENTIAL INDICATOR

It is always advisable to have clear communication when addressing stakeholders to evaluate a proposition and decide if it should be taken to the Execute phase and be launched as the company's next innovation bet. The *Innovation Potential Indicator* is a tangible tool for providing a clear and comparable overview of the explored ideas utilizing the strength of the five lenses of innovation and the company strategy. Start by estimating how much potential the tested ideas have to give clarity and direction. Then score the level of fit across the remaining four lenses, using the same approach as in the previous chapter when desirability was scored.

THE FIT SCALE

Use this scale to evaluate how well the overall concept[1] aligns with a specific lens. The **internal** concept one-pagers are very useful in this assessment. Each level indicates the strength of the fit and helps assess the innovation potential.

Low Fit = 25
The concept does not meet key expectations or requirements. Significant misalignment is present.

Moderate Fit = 50
Some promising elements are present, but some clear gaps or concerns need to be addressed.

DOI: 10.1201/9781003619352-15

This chapter was refined for grammar and fluency using ChatGPT-5.0.

Strong Fit = 75

The concept meets most expectations and shows clear alignment with goals or criteria.

Excellent Fit = 100

The concept fits exceptionally well! No major concerns or changes needed.

Rate the level to the best of your team's knowledge. The knowledge acquired throughout the innovation process, with all the data collected to validate it. Still, be mindful of unconscious bias, and that it is ultimately an assessment including objective, subjective, and predictive views of the idea and fit. With that assessment in mind, it is possible to add a factor to boost or decrease the level of fit if there are certain factors that are important in the specific project scope or the company strategy in general. The lenses and factors can be structured as illustrated in Table 13.1.

The value of each lens is calculated by multiplying the value of a lens by the chosen factor and then dividing the result by the total number of lenses, which is 5.

The formula is:

$$S = \Sigma(L \times F)/5$$

TABLE 13.1

Lens and Factor Overview

Lens	Factor	Explanation
Desirability	Always 1.0	The desirability is based on the idea fits the consumer's need. If the idea doesn't score 75 or 100, it shouldn't still be in scope to proceed forward. A score of 50 needs to be taken back to the process where more understanding is required.
Feasibility	Normal (no or small investment) = 1.0 Willing to invest or use TPM = 1.5	If the product idea is not feasible to produce, but there is a willingness to invest in the factory to make it feasible, use a higher factor than 1.0. It could also be a willingness to use a third-party manufacturer (TPM) to produce it.
Viability	Normal = 1.0 Long-term 1.5	The viability is closely connected to the feasibility. Indicate here, if it is a long-term project with a longer horizon on the return on investment (ROI), use a higher factor than 1.0.
Sustainability	Very important = 0.5 Important = 1.0 Not so important = 1.5	Indicate how important it is that the concept will deliver on the sustainability agenda. If very important, use a score below 1.0.
Responsibility	Very important = 0.5 Important = 1.0 Not so important = 1.5	Indicate how important it is that the concept will deliver on the responsibility agenda. If very important, use a score below 1.0.

LENS	VALUE	X FACTOR	= WEIGHT	/5 →
Desirability	100	1.0	100	20
Feasibility	50	1.5	75	15
Viability	50	1.5	75	15
Sustainability	75	1.0	75	15
Responsibility	75	0.5	38	8
			TOTAL =	73

INNOVATION SCORE

FIGURE 13.1   Innovation Potential Indicator.

S = Innovation Score
L = Value of each lens
F = Factor for each lens

The sum of all weighted values is divided by 5 (since there are five lenses).

As illustrated in Figure 13.1, this will give you a comparable innovation score on which you can qualify your decision on what idea to build a business case on. Consider all the ideas, have a final discussion, and choose which concept(s) to present as a business case in the last decision gate meeting. Show all the tested and evaluated concepts as headlines in a one-pager, together with the associated innovation scores. If there is more than one concept with a high potential and the concepts don't overlap, it can be considered to present both. It will, however, rarely be advisable to present two concepts in one meeting.

DECISION GATE 3

The final Gate meeting determines whether the most promising concept will be approved and moved into the Execution phase. Show the tested concept overview and the assessed innovation potential, and then proceed to the chosen concept. Alternatively, the team may choose to present other high-scoring concepts from the project, but be mindful not to dilute the message of the selected one. There are strong reasons and significant effort behind that choice. It deserves the attention and focus it has earned.

If the chosen concept doesn't gain the necessary support, alternatives can be brought back into play. In some cases, none of the concepts may be signed off on, and the entire project can be stopped altogether. This is never an easy call, especially with all the time, energy, and resources invested, but if moving forward feels too risky, it is better to pull the brake than to use additional resources.

Let's assume this won't be the case and stay optimistic! When combining design thinking with a solid research foundation and a data-driven approach, the chances of presenting a business case with limited desirability or viability are very low. Investments and risk-taking in innovation will, however, always be mandatory.

Before the Gate meeting, the core team should write a solid business case, based on all the learnings from the project. A business case can be shaped and presented in many ways, and there are several important factors to include. Use the five innovation lenses as a structure to present the reasoning behind the innovation potential score. Insights from the Gate 2 meeting will also be included and can serve as a foundation. For the decision board to evaluate whether a concept is worth investing more resources into, it is essential to get a good overview of the total concept and its potential and implications.

Design thinking is often used in F&B innovation to strengthen **consumer-centricity**, but it remains equally important to maintain strong **business-centricity**. All parameters should be assessed through a holistic business value lens.

STORYTELLING

As a *fly in* to the hard numbers of the business case to ensure the business-centricity, it is a compelling start to walk through the process and explain how the team arrived at the final concept and recommendation.

The Design Thinking Narrative

This is how it could sound for the New Dawn project:

Welcome to the meeting! We began this journey with the **pre-scope** of the innovation project titled "New Dawn." Our goal was to explore how we might leverage the breakfast occasion to create a new type of product in our three core markets: UK, Germany, and The Netherlands. Strategically, this project supports the Golden Dawn brand and its ambition to help consumers live healthier lives. We have worked intensively with the New Dawn project and would like to start with an overview of the best-performing concepts. The outcome of our efforts!

As you can see, the highest-scoring concept is actually called the "New Dawn Smoothie Range," designed to fit the "Busy Morning Fuel" Opportunity Space. We would like to begin with a short story about how we have worked:

Who are the consumers who would truly appreciate our suggested smoothie range? We wanted to **UNDERSTAND** the needs of people with busy lifestyles who still aspire to be healthy, as well as the competitive landscape of the smoothie category better. Some of the key questions that guided us were:

- Who are the potential target audiences?
- What do their mornings look like, especially the *on-the-go* moments?
- What are their behaviors and attitudes toward smoothies today?
- What products are hired today? And why?
- Who are the important smoothie producers and why are they successful?
- Should we respond to the veggie trend we are seeing in competitor offerings?

From there, we moved to the **DEFINE** phase, where we synthesized our findings and framed three clear opportunity spaces to guide the process of generating ideas. Next, we entered the **IDEATE** phase, where we ran creative sessions to generate a broad range of ideas, focusing on different formats, benefits, and product experiences. In the **SELECT** phase, we assessed these ideas, narrowing in on the most promising directions by evaluating consumer appeal, technical feasibility, and strategic fit.

Then we moved into the **PROTOTYPE** phase, where we began to bring the selected concepts to life. We created both internal and external one-pagers and developed physical prototypes in an iteration workshop where we experimented with recipes, formats, and ingredient choices.

In the **TEST** phase, we showed the concepts to consumers, and their feedback helped us refine taste, communication, and relevance. The outcome ensured that we are solving real consumer needs in a meaningful way with certain versions of the ideas towards a certain target audience. Talking about the target audience, we would like to illustrate Career Carrie's morning[2] as it could look with the assistance of the New Dawn Smoothie:

> It is a Monday morning, and Carrie wakes to her 5:30 AM alarm and stretches before slipping out of bed. She drinks her water and heads quietly to the kitchen while the rest of the house is still asleep.

> Normally, she'd go back and forth in her head about whether she has time to eat, but not today!
> She grabs a New Dawn Smoothie from the fridge, twists the cap, and takes a refreshing sip.
> It's quick, vibrant, and tasty, and gives her a small win before the morning rush begins.

> By 6:00 AM, the house is buzzing with kids getting dressed, her partner packing lunches, and e-mails are already lighting up their phones. Carrie moves through it all a bit more calmly than normal.

> At 7:30 AM, she's on the metro sipping the rest of her smoothie, listening to her favorite podcast. But unlike most mornings, she's not distracted from the lack of nutrition because she's already had something filling and energizing.

> By the time she hits her desk at 8:30, she's sharper, more focused, and just a little more in control.

> **It's a small shift, but it leaves her feeling lighter, healthier, and ready to take on the day.**

Business Case

Leaving you with the positive feeling of helping Career Carrie have a better morning, let us now reflect and discuss the potential of this product idea and EVALUATE if we want to take it further and enter the EXECUTE phase.

Start with an introduction similar to the example above, adjusted to fit the content of your innovation project and process. Then, go through the business case by answering general questions like the ones below:

Opportunity space and target audience:

- What is the key need the consumer has that the product will fulfill?
- What do they buy today to solve that need? Why is the suggested product better or equally good?

The product and marketing strategy:

- What is the product idea? Write it as a short pitch or use the external one-pager.
- What is the suggested brand and why does it fit?
- What could the product be named?
- How could the value proposition and claims be communicated on the pack? (Include packaging mockups.)
- Which promotional activities would make sense to initiate?
- What is the risk of cannibalization? (The possibility that consumers will stop buying an existing product in the portfolio, because they switch to the new one.)

Product development (NPD) and research and development (R&D):

- Do you know how to create the recipe? The ingredient composition and process parameters.
- Do you have the technology knowledge, or do you need additional expertise?
- Are there some areas where you would need additional time to learn more?

Production and logistics:

- Do we have the needed production line, and does it have additional capacity?
- Do we need to invest in new equipment to produce the product, and what is an estimation of the cost? Referred to as Capex. (Capital Expenditure)
- Can we source the needed ingredients and packaging at a competitive price and produce it at a viable COGS? (Cost of Goods Sold)
- Does the company have access to the needed storage and logistics?

Business model:

- How will we make money? What is the pricing strategy?
- Include the suggested retail price, the estimated COGS, and the marketing investment of the product.
- Who are the target customers and distribution channels?

Market and competitive landscape:

- What are the key competitor products, and how do they compare to your idea? Include price comparison and product composition differentiators.

TABLE 13.2
Innovation Potential Calculation Table and Score

Market and Sales Forecast

Markets	Countries or Areas
Launch date (Potential)	First quarter of next year (Q1 xxxx)
Retail listings	xx (e.g., xx major retail chains)
Distribution points	xx (e.g., xx stores total)
Estimated sales volume	xx units (e.g., xx M/year by Y3)
Cost of goods sold (COGS)	€xx pr unit
Price positioning (RRP)	€x.xx (vs. competitors at €x.xx)
Net revenue	€xx (e.g., €xx M by Year 3)
Gross margin	xx% (The difference between net revenue and the COGS)

Investment Estimates

Capex required	€xx (e.g., for packaging line upgrades)
Marketing investment (A&P)	€xx over 3 years
ROI (Return on Investment)	Payback expected in xx years

Projected size of the opportunity – 3-5 years:

- What will the estimated sales volume be in the first 3 - 5 years? (If it is possible to attain the market size in volume, it is really good. If not, make a "best estimation.")
- It is useful to add some key financial estimates as illustrated in Table 13.2.

Strategic potential:

- What is the channel potential? (Could we expand into e-commerce, food service?)
- What is the international rollout potential?

Partners:

- Can we as a company do it alone, or will we need a partner?
- Could there be a collaboration with a strong brand to create credibility or faster penetration into a category?
- It could be that a company doesn't have the production capabilities or packaging solution needed and hence will need to look for partners.

What could go wrong:

- What legal issues could come up and need to be investigated further?
- Could it be a challenge to get the estimated listings?
- Any other challenges?

FINAL APPROVAL

Based on all the input, questions, and discussions, the decision board will make an informed decision to pause the project, stop it, or approve it. Let's assume that the business case and storytelling are solid enough for the decision board to approve the project and enter the last phase: **The execute phase**.

NOTES

1 Concept = product or product range idea, including any service or business model attached.
2 Use a sketched storyboard to support the morning story, if possible. It will make it even more relatable.

14 Execute (Do It)

PROJECT MANAGEMENT

In the Execution phase of a project, the project leader takes on a highly active role. Several tracks typically run in parallel, involving NPD, Procurement, Production, Quality department, Supply Chain, Marketing, and Sales. Each of these work streams requires a dedicated task leader. In highly complex projects, a technical project manager may also be needed, as the required skillset and knowledge can differ from those of the assigned innovation project leader.

This phase is all about meeting deadlines and ensuring that every element of the project aligns to deliver the final product on time for the agreed launch date. Unlike earlier phases, missing deadlines at this stage can have the serious consequence of not being able to launch as planned. Communication is critical. The project leader serves as a central point of contact, ensuring that everyone is aligned and that any issues, dependencies, or changes are communicated promptly. This requires the ability to make quick and informed decisions, especially when unplanned challenges arise.

As multiple moving parts come together, attention to detail becomes increasingly important. Overlooking even small aspects such as labeling, compliance, shelf-life test, or packaging specifications can result in delays or legal issues. To keep a clear overview and ensure all necessary actions are completed, the project leader must apply a strict and structured approach to coordination. A normal tool for enabling the structured overview is a *Critical Path Timeline*. Some companies use software solutions, but it is also possible to use the good old spreadsheet option, like roughly illustrated in Table 14.1. All companies will use slightly different terminologies and ways of working.

TARGET BRIEF

The target brief is the document where you write the product characteristics that are needed to fulfill the consumer expectation and create a great product experience. Depending on the type of product your company is producing, the product developer will need to know the specific product parameters to develop. If the prototype work has been done properly, a lot of the design parameters will already be written down. Most likely, the product developer has already been deeply involved in the

DOI: 10.1201/9781003619352-16

239

TABLE 14.1
Critical Path Timeline

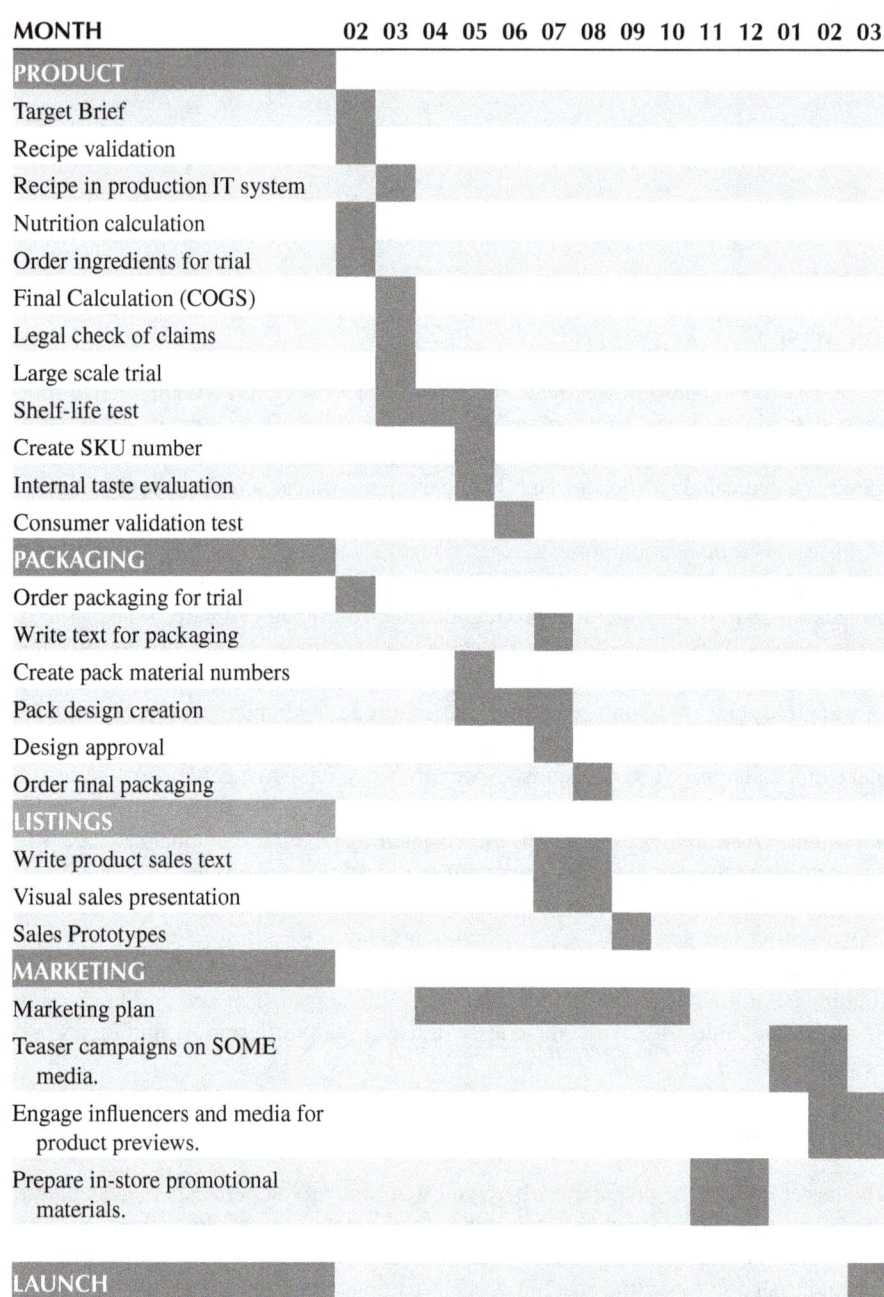

MONTH	02	03	04	05	06	07	08	09	10	11	12	01	02	03
PRODUCT														
Target Brief	■													
Recipe validation	■													
Recipe in production IT system		■												
Nutrition calculation	■													
Order ingredients for trial		■												
Final Calculation (COGS)			■											
Legal check of claims			■											
Large scale trial			■											
Shelf-life test			■											
Create SKU number				■										
Internal taste evaluation				■										
Consumer validation test					■									
PACKAGING														
Order packaging for trial	■													
Write text for packaging					■									
Create pack material numbers					■									
Pack design creation					■									
Design approval					■									
Order final packaging						■								
LISTINGS														
Write product sales text						■								
Visual sales presentation						■								
Sales Prototypes						■								
MARKETING														
Marketing plan				■	■	■	■	■	■					
Teaser campaigns on SOME media.											■			
Engage influencers and media for product previews.												■	■	
Prepare in-store promotional materials.										■				
LAUNCH														■

prototyping, but in larger companies, the final development could also be allocated to another product developer. In this case, it is essential to have a written version of the target brief.

Hereby, an example of points to include in a **target brief:**

PRODUCT DESCRIPTION / TARGET AUDIENCE / USAGE CONTEXT

Use the internal one-pager, potentially a revised version.

Sensory characteristics:

Provide a prototype aligned with the sensory target and relevant competitor products, perhaps identified in the consensus mapping.

Include a description of the following:

- Taste profile in terms of basic taste balance.
- The selected flavor profile(s).
- Texture and mouthfeel.
- Other organoleptic properties as appearance, color, et cetera.
- Serving temperature as chilled, ambient, heated, et cetera.

Nutritional targets:

- Energy (calories per serving or per 100g/ml.
- Macronutrients:
  - Protein: Include target protein content, especially important for products positioned around fitness, satiety, or plant-based alternatives.
  - Fat: Specify total fat and saturated fat.
  - Carbohydrates: Include total carbohydrates and "of which sugars."
  - Total sugar and added sugar: Define limits based on positioning (e.g., low sugar, no added sugar, reduced sugar).
  - Salt level.
  - Fibre: Identify dietary fibre levels if relevant.
- Micronutrients:
  - Vitamins and minerals such as Vitamin D, Calcium, Iron, B12, Potassium, et cetera. (Include whether these nutrients should be naturally occurring or added.)

Ingredient and recipe parameters:

- Key ingredients that must be included due to their functional, nutritional, or storytelling role in the product (e.g., oats, spinach, cold-pressed fruit, et cetera).
- Acceptable ingredients. Specify a list of acceptable ingredients that are permitted for formulation. This may include ingredients that are minimally processed, naturally derived, or approved by specific standards (e.g., organic, non-GMO, EU/US clean label lists, et cetera.)
- Not acceptable ingredients. Clearly outline restricted or prohibited ingredients, which must not be used. These could include artificial sweeteners, synthetic

preservatives, artificial colors or flavors, palm oil, emulsifiers, or other additives that don't align with the product's intended perception of being *natural*, *clean*, or *simple*.

- Sourcing and origin. List if it is important or mandatory to use organic, fair trade, locally produced, et cetera.

Physical and Technical Specifications (Depending on Product Type)
- Target pH level.
- Dry matter content.
- Brix value. (a measure of soluble solids, mainly sugars, in a liquid)
- Water activity. (aw)
- Viscosity. (if applicable, e.g., for sauces, drinks, or purees)
- Density or specific gravity.
- Texture measurements. (e.g., firmness, crispness via instrumental testing)
- Other relevant product-specific measurements. (e.g., carbonation level in beverages, meltability in cheese, et cetera)

Format and Packaging
- Product format. (e.g., single-serve, family pack, multipack, et cetera)
- Packaging type. (e.g., bottle, pouch, tray, et cetera)
- Packaging requirements. (e.g., recyclable, resealable, shelf-ready, et cetera)
- Information about portion size. (e.g., 300 ML, 500 g, et cetera)

Shelf Life and Storage
- Target shelf life.
- Storage conditions. (e.g., ambient, chilled, frozen)
- Handling instructions. (before and after opening)

Preparation and Consumption
- Preparation required. (e.g., ready to drink, heat in microwave, use recipe, et cetera)

Regulatory and Compliance
- Legal requirements in target markets, besides what is stated elsewhere.
- Labeling guidelines. (e.g., allergens, nutrition facts, claims, et cetera)
- Certifications required. (e.g., organic, kosher, halal, et cetera)

Sustainability and Ethical Considerations
- Sustainability goals. (e.g., reduced packaging, low carbon footprint, et cetera)
- Ethical sourcing and social responsibility standards.

Pricing
- Target price point (the COGS).

RECIPE DEVELOPMENT (NPD)

Based on the selected prototype recipe to execute and the information in the final NPD target brief, the product developer can finalize the recipe. Final sourcing of the needed ingredients is essential. Be sure to check availability and price. Use the spreadsheet introduced in the Prototype phase or a recipe management system to create the recipe. Ensure you include all the necessary information, such as nutritional values. Basically, all the essential data needed for ongoing calculations to ensure the recipe meets the requirements and wishes outlined in the target brief. The process parameters are also important to fine-tune and the product developer will need to start planning when to run the production scale-up trials.

UPSCALE AND VALIDATE

When all the work on creating the recipe is done, the time has come to test the product on the production line. The best case is the actual production line that is intended to be allocated when the product is launched, but if that is not possible, the line that will be the closest match. At this stage, it is essential to test wheter the ingredients and process parameters provide the intended product. This can normally be done using the internal expertise in a company or, if needed, external collaboration partners.

Here is a list of some of the important parameters to check:

- Taste: Is the balance of the basic tastes as expected?
- Flavor: Check the flavor profile balance and intensity, including detection of off notes.
- Aroma: Is the smell of the product as expected?
- Color: Are the colors of the product as expected?
- Texture: It can be different parameters depending on the product type. It could be checking intended viscosity, crispiness, smoothness, chewiness, et cetera.
- Food safety parameters: Reach of required PH level, dry matter, et cetera.
- Nutrition: Are the calculated values in accordance with the measured?
- The packaging: A diverse set of functionality aspects, such as whether the packaging contains the product without leaking, what the appearance is, et cetera.

Furthermore, validate that all the raw materials can be sourced in the needed quantities and at the expected price. Because it can influence the viability of the business case.

If everything is in order, these two additional tests are very important:

1. Shelf-life test in a controlled environment.
2. Validation test (CLT).

SHELF LIFE

The shelf life refers to the length of time a food or beverage product remains safe to consume and retains its desired sensory, nutritional, and functional qualities under specified storage conditions. Put another way, a biscuit might still be safe to eat after the best-before date, but if it has gone completely soft and developed an unappealing off-taste from oil oxidation, it should still be considered "too old."

With the increased focus on food waste, it has become a topic that has been debated more in recent years. In Denmark, at least, many products have had additional information added regarding the shelf life. Before, it said: "Best before" or "Last day of sales." Now, the trend is to write: "best before, often good after." A statement that leaves it up to the consumer to assess the product. Both in terms of microbiological safety and the taste experience.

Consumers will rely on appearance and smell. Just as they assess if a product is still good after opening. F&B companies are still responsible for the safety and product experience during the best-before date and normal use, storage, et cetera. To ensure that, they rely on testing and data.

As we touched upon in Chapter 2, *Food and Beverages*, F&B can be divided into storage conditions:

- Chilled
- Frozen
- Ambient

Each of these storage conditions heavily influences how to approach the product design and what they will be able to deliver regarding product experience.

Chilled

Chilled is the most difficult storage condition to handle in terms of food safety. It is the product types with the shortest shelf life. From days to a few months. As we have seen in the New Dawn smoothie project, it is vital to have as long shelf life as possible to reach markets outside of the country of production. With fresh products, limited or gentle processes, this is challenging.

The aim of the New Dawn Smoothie is to create a product experience that is close to how a homemade smoothie would look and taste. A homemade smoothie doesn't last long in the fridge, mainly because of natural processes. They are made with fresh, raw ingredients, rich in natural sugars, moisture, and nutrients. Ideal conditions for the growth of bacteria, yeasts, and other microbes. Even when stored in the fridge these microbes can still multiply over time, just more slowly than compared to room temperature. This gradual activity can lead to changes in texture, sour smells, off-flavors, separation, and potentially harmful bacteria, all part of natural spoilage, especially when oxygen is present.

In the business of producing fresh RTD smoothies, certain measures are taken to prolong shelf life and ensure that no spoilage microorganisms or potential pathogenic bacteria survive or grow during storage. One key measure is pH monitoring, ensuring that the final smoothie consistently has a pH below 4.5. Most likely around 3.50 - 3.90. The product is filled and sealed in pre-sanitized bottles immediately after blending to minimize the risk of contamination.

At the Golden Juice Company, they also apply High Pressure Processing (HPP) to inactivate harmful microorganisms and spoilage bacteria without the use of heat. After production, the smoothies are stored chilled, and microbiological testing is carried out on multiple batches at Day 0, Day 7, and then at weekly intervals until the end of the predicted shelf life, to verify both safety and spoilage resistance.

Standard microbiological tests include the following:

- Total Plate Count (TPC)
- Yeasts and Molds
- Lactic Acid Bacteria
- Coliforms
- Listeria, Salmonella, E. coli

This is just one example of how a chilled RTD smoothie can be produced to ensure it's both safe and delicious to consume. Of course, there is a wide range of other chilled F&B products other than smoothies on the market. Ready meals, desserts, sauces, et cetera. Each with its own set of critical parameters to consider and monitor. Managing these details is all part of the specialized knowledge held by production companies.

Frozen

Frozen is easier and in many cases it enables preservation of the fresh taste of a product. The shelf life can be several months to years. It is good for exporting opportunities.

As an example, ready meal producers can extend their reach with frozen storage. In recent years, there has been a significant trend toward producing Asian food locally. In Thailand, for example, the *Thai Cube*[1] has been a great sales success in the Scandinavian countries and is shipped in large containers. The setup is a small paper box with two compartments. In the big compartment, the steamed rice, and on top, the Thai curry. After preparation, the dish is frozen in highly specialized freeze tunnels, and the shelf life is preserved along with the fresh taste and nutrition. The microwave oven technology brings the dish to life in only seven minutes, and the authentic taste of the incredible food culture of Thailand is conveniently available for global consumers in a safe-to-consume product. Before the products are frozen, extensive microbiological testing is, of course, also carried out.

Ambient

Ambient products cover a very wide range of product types and preservation techniques. It can be products with a high water content, like ambient beverages, canned and jarred food, et cetera. Products with a lower water content, like bread and dried fruit, and the dry products like flour, pasta, spices, et cetera.

Ambient products may come with their own challenges, but they typically offer a long shelf life. Ranging from several months to even years. This makes them ideal for export opportunities as well. To highlight the complexity that can be involved in developing ambient products, granola is a great example to dive into. It is a combined product of different ingredients that all need to fulfill the same ambient storage conditions and shelf life. As an example, a Cereal Company could have a Berry Granola with this taste description:

Our berry granola is a beautiful mix of zingy raspberries and blackberries, crunchy oats and green pumpkin seeds.

The taste experience they aspire to provide is a fresh-tasting granola, as if you had added fresh raspberry and blackcurrant to the traditional mix of oat flakes and pumpkin seeds.

The assumption is that the consumer appreciates fresh berries in combination with their breakfast. It could be a yoghurt that they would want to top with a sweet, crunchy element like a granola and some nice berries. For the consumer, fresh berries are expensive to buy, and they also have a very limited shelf life. These barriers could present a business opportunity to embed the same product experience directly into the granola.

"The benefit of getting the product experience of fresh berries into an ambient product with a long shelf life."

The key challenge here is that the fresh taste of a berry is usually only found in the fresh version. So, what have they done to recreate that same product experience?

These are the ingredients they have used:

Ingredients:

Wholegrain oat flakes (67%), golden syrup (partially inverted sugar syrup), vegetable oil, pumpkin seeds (3.5%), freeze-dried raspberry pieces (1.5%), raspberry purée (1.5%), freeze-dried blackcurrants (1%), blackberry purée (0.5%), natural flavoring.

The first four ingredients are the basic granola part: Oat flakes and pumpkin seeds mixed with oil and syrup, and roasted crispy. The granola base is then mixed with three fruity elements: freeze-dried berry pieces, dried purée, and natural flavoring.

Freeze drying is a very efficient way to preserve the fresh taste of a berry. It can also provide a good red color and some texture from the pieces. The berry taste is most probably boosted with the natural flavoring component. In this way, it is possible to create a "fresh berry" experience in the granola.

Watch outs:

Freeze-dried products are very vulnerable to moisture, and will get soft very easily. It requires special storage and handling at the factory. Special package foil with a high moisture barrier. After the consumer opens the pack, the crispiness of the freeze-dried berries will not last long.

These were some examples of how the product design is deeply related to the shelf life and product experience of products in all three storage conditions. Let's move on with the work still to be done.

To finalize the validation of product quality, it is standard procedure in larger F&B companies to test large-scale production samples with the target audience. This ensures that the taste experience is consistent with the prototypes and that the product still resonates with consumers. This is done by using the CLT test as explained in the Test phase.

It is the final "emergency brake" to be pulled if something is off. Many things can go wrong, but if the design process has been followed and issues have been addressed in good time, many things will go very well! If there are production issues during the first large-scale trial, resolve the issues ASAP and initiate a new one. This time, the products will be in the final type of packaging, without the design.

The work stream of creating the packaging design is ongoing in parallel with the product development. The aim is to have packaging design mockups ready for the CLT test as well, to be able to check how the expectations created by the design align with the product experience.

PACKAGING DESIGN

In parallel with the full-scale trial(s), the work on creating the design of the pack also needs to be finalized and done. As covered earlier in the Prototype phase, packaging plays a critical role in setting sensory and brand expectations through visual cues. All of the same considerations apply here in the Execute phase, but now the design work becomes more refined and concrete. Most F&B companies will hire external support to develop the final packaging, and established companies often have long-standing partnerships with design agencies that handle not only packaging, but also related marketing campaigns. Some companies even have internal design and marketing teams for this purpose.

At this stage, multiple design directions are typically created, drawing on insights gained earlier in the process. These designs should be tested and validated before final decisions are made. While some of the research methods, like conjoint analysis, eye tracking, and focus groups, were already introduced in the Prototype phase, they remain just as relevant now. However, the focus shifts from exploratory insight to confirming what works best.

Other common validation methods include the following:

- A/B testing: Different packaging versions are tested in real or simulated purchase environments to measure actual consumer behavior and preferences.
- Shelf testing: Helps assess shelf standout and visibility in a competitive context.
- Online surveys: Useful for gathering structured feedback on clarity, appeal, and messaging hierarchy.
- In-store pilots: Small-scale launches in selected stores can reveal actual sales performance and shopper feedback.
- Social media testing: Sharing packaging variants online can be a fast and cost-effective way to gauge consumer reactions and engagement levels.

In the New Dawn project, a new brand with a modern identity tailored to the defined target audience was developed.

The Golden Juice Company had a wish to differentiate the new brand from their more traditional Golden Dawn brand. The new range of smoothies for a fast, tiny breakfast at home or *on-the-go* occasions should communicate a fresh new start for both the company and the consumer. The brand name is always a difficult one to align on, but as a general rule, be sure that it is "short and sweet." In the Golden Juice Company they were, however, not in doubt!

The project name "New Dawn" actually captures the essence of what the brand aspired to embody, which was also confirmed with consumers. A brand promise was also set in collaboration with an external agency to help guide the company marketers on how to position and think about the brand. A brand promise can be a few words that capture the heart and soul of a brand: **New Dawn is always: Tasty – Healthy – Natural.**

It is the core promise to the consumers of what to expect when they buy a New Dawn product.

A PET plastic bottle was chosen due to market standards, production requirements, and its lightweight nature, which is important for making global distribution more viable and sustainable.

THE DESIGN

Let's take a look at some of the final design elements that could make it through the "eye of the needle." It is always a balance of what and how much to include … Out of a range of four different smoothies, Figure 14.1 shows how some of these elements could look in the green smoothie version.

FIGURE 14.1   New Dawn Smoothie – Packaging Mockup Example.

PRODUCT EXPECTATIONS

Usage Occasion and Target Audience Fit

- The brand name "New Dawn" and the slogan "A delicious start to the morning" set the occasion for the consumer.
- The design is a combination of modern and traditional elements, providing both coolness and trust. Conveying the brand promise, it fits the target audience by creating expectations that align with their identified lifestyle, needs, and motivations.

Natural Positioning

- The visible product appearance and color build consumer expectations towards naturalness and freshness.
- Cold-pressed is building on the trend of consumers and influencers using the new cold-pressed juicers to make colorful and healthy juices. It conveys that this is also an unprocessed, natural, and healthy beverage.
- The muted mustard background color, RGB: R203, G182, B139, is warm and earthy, evoking natural associations with fruits and vegetables grown in soil.
- Plastic bottles are not normally associated with naturalness and sustainability, and it needs to be called out on the back how to recycle them, perhaps with additional information about the choice of using PET, compared to glass. It is a complex area to communicate, however, as there are many pros and cons to consider.

Flavor and Taste

- **Delicious**, in the slogan, cues great taste.
- The words **vibrant** and **green** build expectations to both taste and health.

Claims and Functionality

- Calling out the specific vitamins, minerals, and fibre inside is not really a claim, as long as the specific amounts are to be found back-of-pack in the nutrition declaration. It leaves it up to the consumer to associate health benefits.
- The logo cues science in a subtle way.

N.B. A word like energizing could be a valuable addition to enhance the expectations; however, some words will be in a grey zone and could be considered a claim. If *energizing* was used as a lifestyle descriptor, as part of branding or mood-setting language, and not linked to specific nutrients or effects, it may not be treated as a formal health claim. But authorities could still challenge it if it misleads consumers. A more significant claim would be something like: Vitamin C – Helps Reduce Tiredness.

Claims like these can be used if approved by the authorities.

Cultural Identity

- Writing Bramley Apples emphasizes where the apples originate. It is a very well-known apple in the UK, not in the other markets, however.
- The company brand logo at the bottom, accompanied by the "Heritage Range" statement, establishes UK origin and conveys premium quality.

THE MARKETING CAMPAIGN

TARGET AUDIENCE

The importance of empathy and truly understanding who your consumers are and why they would buy and enjoy the outcome of your innovation project is just as crucial in the communication beyond the packaging. It is about your marketing efforts to let consumers know that your brand and a new exciting product exist.

Branding a Product is An Art Form in Itself

The starting point in a specific innovation project can vary. Creating a new brand in a start-up environment is often closely related to the product idea. Sometimes it is an existing company wanting to create an additional brand, like the New Dawn example. It is, however, also often the case that marketing efforts are not related to an innovation project and a new launch, but maintaining the awareness and relevance of the current product assortment within a brand. Brands operate with annual marketing calendars, continuously refreshing and fine-tuning both their content and its placement. They must adapt to shifts in consumer behavior and the rise of new platforms. What is "a must" one year, may no longer be a priority the next.

The focus of this book is obviously in connection with an innovation project, and hence, the more long-term work with the marketing efforts of a brand will not be covered. The topics and ways of working involved in marketing and creating memorable brands are vast, and there is an abundance of books written on the topic.

Planning a marketing campaign for a new product launch involves coordinating several elements. The Brand Manager will most likely hire a creative agency if no in-house option is available. In this part of the process, an ideation workshop will also make sense to initiate. In case the creative partner has been involved in the earlier phases of the project and is in tune with the opportunity spaces, it is just a matter of a kick off. If not, they need to be introduced to the specific constellation of demand and the correlations to the created opportunity spaces. They also need to build empathy with the target audience's needs and motivations to enable their creativity. In essence, it is about clearly defining who the campaign should target – both on an emotional and functional level. Where and when to engage with consumers, and how to tell the story in a way that builds awareness and sparks interest. That is why it makes sense that the marketing department and the creative agency(s) working on both the packaging design and the marketing campaign, should be involved in the consumer research from the start.

As Danielle Sarver Coombs so precisely frames the consumer perspective on brands and their products, in her brilliant consumer insight handbook:

The brands gets you, so you get the brand[2]

Back to the New Dawn project, where it is a case of both launching a new range of smoothie products and simultaneously creating a brand. Previously, the team defined the brand promise: Tasty – Healthy – Natural. The next important task is to bring the brand promise to life with a consistent narrative that is catchy and can be built on.

This narrative is called a creative platform[3] or the "big idea." As an example, Andrew Geoghegan[4] points out that the creative platform for Snickers' advertising is the well-known campaign titled: "You're Not You When You're Hungry." This campaign is built on the relatable insight that hunger can lead to noticeable personality changes. The campaign humorously portrays individuals acting out of character due to hunger, only to return to their usual selves after consuming a Snickers bar.

The point is that the "big idea" is strong enough to endure over time, allowing for fresh, creative campaigns built on the same resonant foundation. While the Snickers campaign has evolved through the years, its core idea has remained consistent: Hunger causes noticeable changes in a person's behavior or identity. The tone has stayed humorous and exaggerated, always ending with the resolution that eating a Snickers brings things back to normal. There are no set rules on how to do this and what sparks the "big idea." It is a genuine creative exercise, getting inspired by all kinds of sources like humor, movies, science, art, poetry, ordinary lifestyle – or the extraordinary, sport, compulsive behavior, music, et cetera.

Imagine the New Dawn creative platform as a fresh take on the iconic song *Feeling Good*[5] with its memorable lines:

It's a New Dawn
It's a new day
It's a new life
For me
And I'm feeling good …

It can set the tone and rhythm for a campaign, serving as a catchy musical backdrop, while sparking a wealth of creative ideas on how to tell the remarkable and contagious story of a new smoothie range and brand hitting the market: A New Dawn is coming. A new life even! – a new vibrant smoothie that will leave you feeling good …?

I will leave it up to you to imagine how …

CHANNELS

The next step is to understand the best time and place to connect with your audience and share the story about the new product and brand. At this stage, it is valuable if consumer research explored where the target audience gets their information and inspiration. Think back on the consumer journey and imagine how they interacted with friends and family, sharing recommendations during dinner parties, lunch breaks, or casual conversations, and discovered interesting food and beverage products and brands. Where else did the information come from? How much information do they get from social media channels like Instagram, Facebook, TikTok, YouTube, Pinterest, Snapchat, or whatever platform is prevalent in a specific country? Which is the preferred? – keeping in mind that today's trending platform might be replaced by something new tomorrow.

When do they interact with these platforms, and do they also use other sources for inspiration like food blogs, TV cooking shows, recipe magazines, influencer reviews, or podcasts? And to what extent do they notice in-store advertisements, outdoor billboards, and so on? Do they appreciate branded events, product placement, or pop-ups at festivals, where they can try new products and feel the brand vibe?

ACTIVITIES

With a clear understanding of the channels the target audience frequently uses or is exposed to, begin planning the specific activities for the product launch marketing campaign. There is an abundance of potential activities to choose from, which once again can be categorized as digital, analog, or hybrid. Here are some examples:

Potential Digital Marketing

Content Marketing
Creating and sharing useful and interesting content on relevant platforms to build trust by offering value, without asking for anything in return. It provides the foundation for other marketing efforts by giving audiences something meaningful to engage with.

Social Media Campaigns
Promoting content, products, or services through social media platforms helps increase brand awareness, boost engagement, and drive traffic or prompt actions such as sign-ups, downloads, or purchases.

Video Marketing
Using video content across TV, streaming platforms, and social media to explain, entertain, or showcase products and services in ways that capture attention and engage audiences visually.

Influencer Marketing
Partnering with influencers or content creators to promote your brand or product to their followers authentically and engagingly.

Email Marketing
Sending personalized and timely emails to engage subscribers, build relationships, share updates, and promote relevant offers. Although it may sound somewhat

old-fashioned, it continues to be an effective channel for maintaining ongoing consumer engagement.

Search Engine Optimization and Marketing

Improving website content and structure to increase visibility in search results, while also running paid ads to target specific keywords and attract relevant audiences actively searching for related information or solutions.

N.B. Marketing to AI Assistants

Given the rapid advancement and growing integration of artificial intelligence into everyday life, it is highly plausible that AI assistants will increasingly function as trusted agents, curating information, generating recommendations, and executing purchases on behalf of users. Consequently, the focus of digital marketing may shift from targeting the end consumer directly to engaging with the AI assistant itself.

Potential Analog Marketing

Potential In-store Promotions

Offering in-store experiences that encourage product trial and visibility. Sampling stations invite consumers to taste the product, especially during busy hours, increasing the chance of spontaneous purchase. Point-of-sale (POS) displays use eye-catching visuals to highlight key benefits such as natural ingredients and convenience.

Potential Traditional Media

Using established media channels to build awareness and reach wider audiences. Printed ads placed in relevant publications help communicate product benefits to the target audience. Billboards and transit advertising with strong visuals and short messages are effective for capturing attention during daily commutes.

Potential Hybrid Marketing

Public Relations

Generating awareness and credibility through both digital and traditional media can include press release announcements. These typically share information about the product benefits, the target audience, and the brand promise. They may also highlight a strategic partnership. For example, a collaboration with a well-known fitness chain could help position a product within a health-conscious lifestyle.

Customer Engagement

Loyalty programs that reward repeat purchases and strengthen brand connection can help build lasting relationships with consumers across platforms.

With all the available knowledge and creative ideas, plan the marketing campaign and timeline. A launch plan could look something like this:

LAUNCH PLAN

Pre-Launch

- Run teaser campaigns on social media.
- Engage influencers and media for product previews.
- Prepare in-store promotional materials.

Launch

- Activate in-store promotions and sampling.
- Launch the full digital marketing campaign.
- Roll out bus banner ads and platform advertising in subways and trains.
- Distribute press releases and engage with media outlets.

Post-Launch

- Monitor sales and collect consumer and customer feedback.
- Adjust promotional strategies based on performance.
- Introduce loyalty programs and initiatives to drive ongoing engagement.

The launch plan must be developed in close collaboration with the core team, especially sales, to ensure alignment on campaign activities, potential listings, and store availability. Few things are more frustrating than seeing an advertisement for an exciting new product, deciding to try it, and then not finding it in stores! With that in mind, let's look at how to enable sales teams to get the essential listings. Failing to get sufficient listings will, after all, result in a very critical situation. Without them, the launch risks falling short, reducing the entire effort to an expensive learning experience.

GET LISTINGS

During the Execute phase, it is also time to get the listings at the customers in place. It is a good idea to provide the company's sales employees with full-scale samples and compelling concept material to enable them to make the product attractive.

There are several important areas to be able to communicate with the retailers:

A. How well will the new product fit **consumer** needs and motivations?

Ensure describing a target audience that doesn't sound like too small a segment of the population. Provide the storytelling about the innovation project and the journey to the new product. Who it is for and why they will find it a product worth trying. Why they will re-buy it and start advocating for it? In case it is also a new brand, what is the brand promise and why will the consumers love that? It is always easier to sell in a well-known brand that is liked and trusted, compared to a new one.

B. How well will the new product fit **customers' own** needs and motivations?

They are, of course, interested in a new product that the consumers would love to buy, covered by the first point, but retailers also have additional concerns. Will they have a good margin on the sales? That's key, but far from the only consideration.

Retail procurement teams also look at the category fit. Does the product complement or expand the existing range in a way that makes sense for them? Retailers like innovative products, but they must fit within their category logic. They might ask: If this product is going on the shelf, what product is going out? – as we can't allocate more shelf space in this category. And the most obvious product might also be provided by your company, and where will that leave the business case? Retailers prefer products that are easy to store, have a good shelf life, and show strong rotation, as this helps keep shelves efficient and avoids waste.

They will also consider whether the product is priced in line with consumer expectations and will ask about the plans to support it through in-store promotions and discounts. If your company is new to the retailer, they will also ask about supply reliability and logistics. Is your company integrated into existing distribution systems, and will you be able to consistently deliver the product in the right quantities, on time, and without quality issues? Retailer and supplier relationships are a matter of trust and take effort and time to build.

Finally, it is a powerful argument to visualize how the packaging stands out on the shelf and is easy for consumers to understand.

In addition to physical mockups and product samples, have visualizations ready showing what the new products would look like on the shelf. This makes it tangible for the buyer and helps them picture the in-store impact. If you also have consumer research showing a preference for the design, your case becomes even stronger. To put it simply: great shelf appeal drives purchase, and strong purchase potential drives **procurement appeal**.

READY TO LAUNCH

If everything goes according to plan and the critical timeline is kept, the entire innovation process will come to an end. By implementing design thinking methodologies, the outcome will be strong. The product delivers on taste and functionality, tailored to fit a well-researched opportunity space. It comes in a packaging with clear communication to the consumer, creating the desired set of expectations to be fulfilled in the product experience. The packaging has remarkable appeal in-store and during usage. The product is supported by a genuine and cool brand and a captivating marketing campaign. The entire product concept has been evaluated and balanced within the five lenses of innovation. A collaborative innovation project delivered a product using creativity enabled by empathy. And actually, not just consumer empathy – a holistic empathy covering consumer, customer, own business, and society.

LET'S LAUNCH THIS PRODUCT!

NOTES

1   www.kitchen-joy.com/
2   *The Consumer Insights Handbook: Unlocking Audience Research Methods*, Danielle Sarver Coombs, Rowman & Littlefield Publishers, London, UK, 2021. Page 1.

3   *Effective Brand Building: Unlock Growth with Strategy, Insights and Measurement,* Andrew Geoghegan, Kogan Page, London, UK, 2025.

4   *Effective Brand Building*, Geoghegan (2025).

5   From the song "Feeling Good," originally written by Anthony Newley and Leslie Bricusse for the 1964 musical The Roar of the Greasepaint – The Smell of the Crowd. The song gained widespread popularity through Nina Simone's 1965 recording, which has since become iconic. (https://en.wikipedia.org/wiki/Feeling_Good)

15 Launch

CELEBRATE!

Any company that goes through all the steps of developing and launching a new product should take some time to celebrate it.

THROW A LAUNCH PARTY

A launch party isn't just a chance to unwind! It is a moment to reflect, recognize, and reconnect as a team. After the intensity of an innovation project, pausing to celebrate the collective effort is crucial for morale and company culture. Set the scene by decorating the space at your company in a way that reflects your product and brand identity. Invite all colleagues and stakeholders – not just those who worked directly on the project. It is a gesture of inclusion, signaling that everyone contributes to making launches possible, whether working in HR, finance, logistics, or beyond. Have samples of the product ready to hand out to everyone, along with custom-made merchandise to mark the occasion – think branded T-shirts, tote bags, or even quirky items tied to the product or brand. Not only does it build pride, but it also becomes a visual reminder of what the company has accomplished. Show the journey with *Behind the Scenes* videos and pictures that capture key moments from the project. Include everything from early sketches to funny outtakes and fieldwork footage. A well-crafted launch party does more than celebrate – it builds emotional connection to the work, the brand, and company spirit.

GIVE IT TIME!

The success of a new product launch is rarely instant. The reality is that real market traction takes months or even years. The consumers have to notice the new product, understand it, try it, and then return to it. That journey takes time.

If possible, have the patience and responsiveness to give your product the time and support it needs to grow. When a company invests the time and effort that an innovation project demands, it is only common sense to support it with strong marketing and sales work. Bearing in mind that retailers and other customers can also decide to

DOI: 10.1201/9781003619352-17

This chapter was refined for grammar and fluency using ChatGPT-5.0.

delist a product again if they don't see the expected value or indications that the sales will grow.

That said, it is also quite common for a new product launch to get off to a great start, exceeding expectations. This success is often driven by initial hype, most likely amplified by a brilliant marketing campaign. The real measure of success comes after consumers experience the product and reflect on whether it met their expectations.

EVALUATE AND LEARN!

The Project Evaluation

After a project and the launch of a product, it is good practice to look at the process and evaluate whether anything could have been done better.

Ask questions like the following:

- Did we allocate our time and resources in the best possible way?
- Did we have the right team?
- What was difficult and why?
- What can we do better in the next project?

The Product Performance Evaluation

At some point in time, it will also be clear whether the launch is performing on sales, and feedback from customers and consumers will also be able to be collected. Some of the key performance indicators (KPIs) of a new product launch, which could be helpful to gather and analyze, are:

The hard figures:

- Market penetration: Measure of the product's availability across targeted retail outlets.
- Sales volume: Track units sold in the first six months post-launch.
- Repeat purchase rate: If possible, track how many customers purchase the product more than once. To access that type of data, a close collaboration with a retailer could be an option, or data from specific consumer surveys.

The feedback:

- Consumer feedback: Scan and analyze reviews and ratings already available online. Initiate consumer surveys asking for feedback.
- Customer feedback: Use the sales team to investigate how the important deliverables of the product, as we touched upon earlier, are assessed by the customers.
- Social media engagement: Monitor likes, shares, comments, and influencer impact.

Start to analyse all these data points and reflect on what your expectations were and why. Especially the points that bring a surprise are valuable to deep dive into. Understand the reasons behind and learn from them for the next project. It is important to learn from the successes and failures in a company to improve the ways of working. Remember to write down the lessons learned and have a cloud drive dedicated to the purpose.

Next time a famous phrase in a future innovation project is uttered:

We already tried this and it didn't work!

Previous failures and successes have been analyzed in a structured and thorough way, enabling you to add why it didn't work.

A more constructive statement like this can be a build:

Yes, we already tried this and it didn't work because of XYZ. However, we could look at it again if the situation has changed and if XYZ is addressed differently in the process or execution.

A much better starting point to evaluate if a previous idea will have current potential to be the answer to an identified consumer challenge, more easily produced, or positioned at a better price, contribute to more sustainable lifestyles, making a difference in a local community, or something else within the five lenses of innovation.

FINAL THOUGHTS

Design thinking for Food and Beverage Innovation is a mindset! A mindset that puts the consumer in front and interacts with them during the innovation process.

This book has been written with the ambition to show how design thinking can be utilized in F&B innovation. Hopefully, it has been supported by relatable and explanatory examples that show how to implement the mindset into the innovation process. A process where data goes hand-in-hand with gut feeling, experience, and expertise, and the outcome is delivered by creativity enabled by empathy with the end user of the innovation, the consumer.

I initially described the aim as creating a *traveler's guide* to navigating the F&B innovation process. A detailed guide on how to start an innovation project and all the avenues of research and tools that can assist you in the journey. It is my hope that, if you have come as far as reading this, the content of the book will have provided exactly that.

It should also be clear that the amount of resources that can be allocated to implementing design thinking in innovation naturally depends on the type and size of the F&B company. Be sensible about what to prioritize and find the balance that fits the constraints on resources, time, and the potential return on investment. Every project is different, and the approach will always need to be tailored to match it. I love the mindset and *hands-on* agility of startups, and I also highly value the more structured and data-based possibilities that larger corporations offer. I believe both cultures can learn a lot from each other.

It is by no means an easy journey to transform existing ways of working within innovation. Whether it is a small or large company setup, in all circumstances it will take time, effort, and valid arguments. Promoting and implementing a design thinking mindset, along with its methodologies, into new food and beverage innovation processes can be as difficult as learning to surf the cold, untamed waves along the Klitmøller coastline on the west coast of Denmark (aka Cold Hawaii). A challenging endeavor, but also great fun and deeply rewarding when successful.

Appendix

WHY B2B COMPANIES SHOULD WORK WITH DESIGN THINKING

It is pretty straightforward why B2C companies should work with consumer centricity and use design thinking methodologies. It makes sense to develop products that the consumers would like to buy.

But why should B2B companies work with design thinking, and why is it rarely the case that they do? As you might remember, my own attempt to change the ways of working and implement parts of design thinking in a B2B company was unsuccessful.

Let's start with the reasons why they don't have this focus. One argument could be that they are not selling directly to the consumers in the same way as B2C. This is, in some ways, true. The consumers will eventually end up consuming their products, but most of the time, it will only be as a part of the ingredient list. The B2B company does not have any control over the "end product" and how it is positioned anyway. It could also be argued that the B2B companies are not set up to work with the design thinking methodology. Also true, but why should they, if they don't work with it? A catch-22 situation …

Mostly, B2B companies allocate their resources primarily toward enhancing the sales force in their interactions with customers. The account manager and perhaps an application specialist who can showcase the great functionality of the ingredients. Price and functionality are key drivers and are essential to prioritize. No doubt about that!

So, how to differentiate yourself as a B2B company?

I can see 2 ways to use design thinking to leverage sales in B2B companies.

CUSTOMER CENTRICITY

Take a deeper look at your customers and use design thinking methodology to better understand their needs. Price and functionality, yes! Price is always key, but what other aspects connected to price and production costs could be of importance? Plug-and-play solutions, reduce complexity, higher yield, et cetera.

Ingredient functionality? Big topic, complex, and a lot to understand!

What questions would a professional researcher ask production line operators to uncover opportunities for improving production efficiency and enhancing product quality? How is the current packaging format working for the customer? When they unpack them, for example. How do they store your product? Would it be possible to produce completely new types of products on the production lines available to the customers? How can the B2B company understand the CSR policy of the B2C company and develop customized solutions that align with their CSR commitments? Even to the extent of looking at the B2C company's waste and suggesting ways to add value to these side streams. The possibilities are many, but of course, this approach also comes with some challenges. How to start and explain to your customers what you are aiming for can be tricky.

I recommend starting with a long-term collaboration partner and having an introductory meeting to explain how you intend to gain a deeper understanding of their production and business. A deeper understanding serves as a foundation for developing innovative, optimized solutions that will benefit them. After a successful project and satisfied customers, the story becomes easier to communicate and can serve as leverage for future projects with other customers.

Address concerns about disclosing company secrets with tailored non-disclosure agreements. Avoid making them too broad. Clear, focused agreements build confidence and ensure protection while preserving the freedom to operate.

CONSUMER CENTRICITY

Some B2B companies offer consumer insights and trend research as a customer benefit. Big flavor companies and other major ingredient suppliers, in particular. These companies often have dedicated consumer insight departments and strong financial resources. They visit their customers to present ingredient trends, consumer occasions, market analyses, and more. This is all valuable input. But do B2C companies know how to make the most of this information? Some do, of course. But many lack the capabilities to turn these insights into meaningful innovation!

The reality is that innovation isn't as simple as receiving a deck of trends or a report on consumer behavior. Insights alone aren't enough. What's often missing is the ability to connect the dots – to translate those insights into clear, compelling product ideas that address real consumer needs. This is where design thinking comes into play. As you have read about, it provides the structure and tools to bridge the gap between insight and action. Could it be the missing piece in the puzzle for B2B companies looking to support their strategically important customers? Helping them turn the insights into relevant solutions for consumers. What if B2B companies started offering full product design packages to their customers?

These could include consumer insights, workshop facilitation, and support with prototyping and testing. Not just data, but a pathway to action. You might argue that this kind of facilitation is already available through the agencies we just covered – and you are right. Their expertise is essential. But imagine if those agencies were brought in by the B2B companies themselves, forming strategic partnerships that span all three players: B2B, Agency, and B2C.

Just a thought to consider …

ROLE VOCABULARY

Companies often use different titles for similar responsibilities, and roles may blend across functions depending on the organization. Titles are not always used consistently, and new ones continue to emerge as innovation processes evolve. Structures and titles vary widely from one company to another, but here are some of the most commonly used terms and roles.

Artwork Designers: Creative professionals who develop visual assets for F&B packaging, including layout, graphics, typography, and labeling elements. Often hired through creative agencies. They ensure that packaging aligns with brand identity, regulatory requirements, and consumer expectations. They collaborate closely with marketing, packaging, and regulatory teams throughout the innovation process.

Brand Manager: A strategic role, responsible for shaping and maintaining a brand's identity, positioning, and performance in the market. They translate consumer insights into brand strategies, oversee product messaging and campaigns, and ensure consistency across touchpoints to build brand equity, drive growth, and increase category penetration.

Category Manager: A commercial role responsible for overseeing the strategy and performance of a specific product category. They balance consumer insights, market dynamics, and business objectives to optimize assortment, pricing, and promotions, working closely with marketing, sales, and innovation teams.

CEO (Chief Executive Officer): The highest-ranking executive in a company, responsible for making major corporate decisions, managing overall operations, and acting as the main point of communication between the board of directors and corporate operations.

Chemistry Food Scientist: A specialist who applies chemical principles to understand and improve the composition, stability, and interactions of ingredients in F&B products. They support product development by analyzing flavor, texture, color, and shelf life, and ensure quality, safety, and regulatory compliance through analytical testing.

Consumer Insights Manager: A specialist who leads the generation and interpretation of consumer research to guide brand, product, and innovation strategies. They manage qualitative and quantitative studies, translating findings into actionable insights that support decision-making and long-term growth.

Consumer Scientist: A specialist who researches consumer behavior, needs, barriers, and motivations. They define and scope segments to uncover insights that guide innovation and marketing strategies.

Culinary Developer (Chef): A hands-on creative role that contributes with culinary expertise to concept development and rapid prototyping. Skilled in combining ingredients and preparation methods, they collaborate with NPD specialists and product designers to shape ideas into feasible, consumer-relevant food experiences. Their practical approach supports fast iteration and early-stage innovation.

Data Analyst: A specialist who collects, analyzes, and interprets data to support decision-making across marketing, sales, innovation, and operations. They turn consumer, market, competitor, and internal sales data into actionable insights, often using predictive models to optimize product launches, category strategies, and business growth.

Innovation Manager: A key role responsible for driving the development of innovations that align with business strategy and consumer needs. They identify opportunity spaces, lead cross-functional teams through the innovation process, and manage pipelines to deliver differentiated products that support growth and category expansion.

IP Specialist: A professional responsible for protecting intellectual property related to F&B innovations. They identify and manage assets such as recipes, processes, packaging designs and trademarks, supporting innovation teams by coordinating patent and trademark filings, assessing risks, and aligning with legal and regulatory requirements.

Key Account Manager (KAM): A commercial role focused on managing and growing strategic customer relationships with priority retail or foodservice accounts. They are responsible for delivering sales targets, aligning joint business plans, and ensuring execution of agreed initiatives. Working closely with marketing and supply chain, they help secure distribution, visibility, and promotions that support brand and category growth.

Licensing Manager: A role focused on managing brand partnerships and intellectual property agreements. They negotiate and oversee licensing deals that bring external brands, characters or concepts into product development, ensuring alignment with brand strategy, legal requirements and commercial goals.

Microbiology Food Scientist: A specialized role that focuses on the study of microorganisms in F&B products. They ensure product safety and quality by assessing microbial risks, validating preservation methods, and supporting shelf-life testing. In some cases, they also contribute to fermentation-based innovation and process development.

Nutrition Food Scientist: A specialist who focuses on the nutritional composition and health impact of F&B products. They work to optimize recipes for nutritional balance, support regulatory compliance for nutrition claims, and contribute to the development of healthier innovations that align with consumer needs and dietary trends.

Open Innovation Manager: A strategic role that identifies and manages opportunities to collaborate with external partners such as startups, universities or suppliers. They scout new technologies, ingredients or business models to support internal innovation pipelines and accelerate product or process development.

Packaging Manager: A specialist responsible for developing and managing packaging solutions that protect F&B products, support brand positioning and meet sustainability, regulatory and operational requirements. They collaborate with design, R&D and supply chain teams to ensure packaging aligns with consumer expectations, production capabilities and environmental goals.

Plant Manager: A senior operational role responsible for the day-to-day performance of a manufacturing facility. They ensure that production runs efficiently, safely and in line with quality and cost targets, while supporting the successful scale-up and launch of new innovations.

Procurement Specialist: A supply chain professional responsible for sourcing ingredients, packaging and other goods or services. They manage supplier relationships, negotiate contracts, and ensure cost efficiency, quality and continuity of supply, while supporting sustainability and compliance goals.

Process Specialist: A technical expert focused on developing and optimizing manufacturing processes. They ensure scalability from pilot to full production, maintain product quality and efficiency, and support innovation by translating product concepts into stable, cost-effective production methods.

Product Designer: A role still emerging in the F&B industry, but increasingly adopted by progressive companies. A key creative role in the conceptualization, design and development of consumer centric innovations, focusing on functionality, taste, packaging design and ensuring alignment between the expectations set by the packaging and the actual product experience.

Product Developer (NPD): A core role in the F&B industry focused on turning concepts into commercially viable products. Responsible for translating ideas into formulations, refining taste and texture, and ensuring products meet technical, nutritional and regulatory requirements while staying aligned with brand positioning and consumer expectations.

Project Manager (Project Leader): The project manager is the central driver of the innovation process, responsible for keeping everything on track and facilitating communication between teams. They make sure that the right people are involved at the right time and that momentum isn't lost. In some companies, the project manager is an independent role, while in others – especially smaller or mid-sized companies, the role may be taken on by the innovation manager or even the brand manager. It depends on team structure, available resources, and the complexity of the project.

Quality Assurance (QA) Manager: A technical role focused on maintaining the safety, consistency and compliance of products and processes. They develop and oversee quality systems, conduct audits and testing, and ensure that innovations meet regulatory and internal standards before and after launch.

Regulatory Affairs Specialist: An expert responsible for ensuring that F&B products comply with local and international regulations. They manage ingredient approvals, labeling requirements, and health or nutrition claims, supporting innovation by navigating regulatory frameworks and minimizing compliance risks.

Sales Director: A senior commercial role responsible for leading the sales strategy and performance across channels and key accounts. They manage the Key Account Managers, ensuring alignment and excellence in execution. The Sales Director drives revenue growth and helps to build strong customer relationships.

Sensory Food Scientist: A specialist who designs and conducts sensory evaluations to understand how consumers perceive F&B products. They assess attributes like taste, aroma, texture and appearance, translating sensory data into insights that guide product development, quality control and consumer preference optimization. They might also lead a panel of professional tasters.

Sustainability Specialist: A professional focused on reducing the environmental and social impact of products, processes and operations. They develop and implement strategies related to sustainable sourcing, packaging, waste reduction and climate goals. Working across functions, they support innovation, compliance and communication of sustainability efforts to stakeholders and consumers.

Supply Chain Manager: A strategic role responsible for overseeing the end-to-end flow of materials, products and information across the value chain. They ensure efficient planning, sourcing, production and distribution while balancing cost, quality, service levels and sustainability to meet consumer and business needs.

Technical Project Manager: A cross-functional coordinator responsible for planning and executing product and process development projects. They manage timelines, resources and technical deliverables, ensuring alignment between R&D, supply chain, marketing and operations to bring innovations to market efficiently and on time.

Workshop Facilitator: A professional who designs and leads collaborative sessions to drive idea generation, problem-solving and alignment across teams. They facilitate workshops with stakeholders from marketing, R&D, consumer insights and beyond, using structured methods to spark creativity, explore opportunities and accelerate innovation processes.

SOURCES

Blue Ocean Strategy: How to Create Uncontested Market Space and Make the Competition Irrelevant (Expanded Edition), W. Chan Kim & Renée Mauborgne, Harvard Business Review Press, Brighton, MA, 2015.

Change by Design, Revised and Updated: How Design Thinking Creates New Alternatives for Business and Society, Tim Brown & Barry Katz, Harper Business, New York, 2019.

Competing Against Luck: The Story of Innovation and Customer Choice, C. M. Christensen, K. Dillon, T. Hall, & D. S. Duncan, Harper Business, New York, 2016.

Consumer Behavior (Fifth edition), Zubin Sethna, Sage, New York, 2023.

Design-Driven Innovation: Changing the Rules of Competition by Radically Innovating What Things Mean, Roberto Verganti, Harvard Business Press, Boston, MA, 2009.

Driving Innovation from Within: A Guide for Internal Entrepreneurs, Kaihan Krippendorff, Columbia University Press, New York, 2019.

Effective Brand Building: Unlock Growth with Strategy, Insights and Measurement, Andrew Geoghegan, Kogan Page, London, 2025.

Fornemmelse for smag, Ole G. Mouritsen & Klavs Styrbæk, Nyt Nordisk Forlag, Copenhagen, 2015.

Hidden in Plain Sight: How to Create Extraordinary Products for Tomorrow's Customers, Jan Chipchase & Simon Steinhardt, Harper Business, New York, 2013.

How to Avoid a Climate Disaster: The Solutions We Have and the Breakthroughs We Need, Bill Gates, Knopf, New York, 2021.

Identity Designed: The Definitive Guide to Visual Branding, David Airey, Rockport Publishers, Gloucester, Massachusetts, 2019.

Koji Alchemy: Rediscovering the Magic of Mold-Based Fermentation, Jeremy Umansky & Rich Shih, Chelsea Green Publishing, Chelsea, Vermont, 2020.

Packaging Design: Successful Product Branding from Concept to Shelf, Marianne R. Klimchuk & Sandra A. Krasovec, Wiley, Hoboken, New Jersey, 2012.

Sensory and Consumer Research in Food Product Design and Development, Howard R. Moskowitz, Jacqueline H. Beckley, & Anna V. A. Resurreccion, Wiley-Blackwell, Hoboken, New Jersey, 2006.

Sprint: How to Solve Big Problems and Test New Ideas in Just Five Days, J. Knapp, J. Zeratsky, & B. Kowitz, Simon & Schuster, New York, 2016.

The Art of Fermentation, Sandor Ellix Katz, Chelsea Green Publishing, Chelsea, Vermont, 2012.

The Biological Basis of Food Perception and Acceptance. Food Quality and Preference, L. M. Bartoshuk, Elsevier, The Netherlands, 1993.

The Consumer Insights Handbook: Unlocking Audience Research Methods, Danielle Sarver Coombs, Rowman & Littlefield Publishers, Maryland, 2021.

The Creative Act: A Way of Being, Rick Rubin, Penguin Press, London, 2023.

The Design Thinking Playbook: Mindful Digital Transformation of Teams, Products, Services, Businesses, and Ecosystems, Michael Lewrick, Patrick Link, & Larry Leifer, Wiley, Hoboken, New Jersey, 2018.

The Design Thinking Toolbox, Michael Lewrick, Patrick Link, & Larry Leifer, Wiley, Hoboken, New Jersey, 2020.

Winning at New Products: Creating Value Through Innovation (5th ed.), Robert G. Cooper, Basic Books, New York, 2023.

Index

For Product Safety Concerns and Information please contact our EU
representative GPSR@taylorandfrancis.com
Taylor & Francis Verlag GmbH, Kaufingerstraße 24, 80331 München, Germany

www.ingramcontent.com/pod-product-compliance
Lightning Source LLC
Chambersburg PA
CBHW060825170526
45158CB00001B/82